Chapter 1: The Science of Oxytocin

The Discovery and History of Oxytocin

The hormone oxytocin, often referred to as the "love hormone," has a rich history of discovery and scientific exploration. First isolated in 1906 by Sir Henry Hallett Dale, oxytocin was initially recognized for its role in stimulating uterine contractions during labor and aiding in milk ejection during lactation. However, it was only in the latter half of the 20th century that researchers began to understand oxytocin's far-reaching effects beyond reproduction, including its impact on emotional bonding, social behavior, and mental health.

Oxytocin's discovery was a pivotal moment in understanding the biochemistry of human behavior. It wasn't until the 1950s that Dr. Vincent du Vigneaud, a biochemist, isolated oxytocin's full molecular structure and synthesized it in the laboratory. This marked the beginning of a new chapter in understanding hormones not only as physical regulators but as essential players in the realm of emotional and psychological health. With the ability to synthesize oxytocin, scientists began to explore its diverse applications, including its use in reproductive medicine and, later, its impact on human behavior.

Basic Biology of Oxytocin and Its Receptors

Oxytocin is a peptide hormone produced in the hypothalamus, a small but critical area of the brain that acts as the body's command center for regulating various vital functions, including hunger, sleep, and temperature. From the hypothalamus, oxytocin is transported to the posterior pituitary gland, where it is released into the bloodstream in response to specific stimuli.

The biological functions of oxytocin are mediated through oxytocin receptors, which are located throughout the body, particularly in the uterus, breasts, brain, and heart. These receptors are protein molecules embedded in the membranes of cells. When oxytocin binds to these receptors, it triggers a series of intracellular events that lead to the hormone's effects, such as uterine contraction during childbirth or milk release during breastfeeding.

Recent research has shown that oxytocin receptors are also located in regions of the brain involved in emotions, decision-making, and social bonding. These receptors are crucial for the hormonal effects that influence behaviors such as trust, empathy, and affiliation, and they help explain oxytocin's broader impact on human psychology and social interactions. Notably, oxytocin receptors have been found in areas of the brain such as the amygdala, hypothalamus, and prefrontal cortex, all of which play key roles in emotional regulation, social recognition, and stress responses.

Oxytocin's Role in Human Physiology

Oxytocin's role in human physiology is multifaceted, as it governs a wide range of functions related to reproduction, social bonding, and emotional regulation. Among its most well-known functions is its involvement in childbirth and lactation.

Childbirth and Labor

Oxytocin is perhaps most famous for its role in facilitating labor. During childbirth, oxytocin is released in large quantities, causing the smooth muscles of the uterus to contract. These contractions help push the baby through the birth canal. In the final stages of labor, oxytocin levels rise dramatically, promoting more frequent and intense contractions. This action, often referred to as "the oxytocin cascade," is essential for a successful delivery. When labor fails to progress or when induced labor is medically necessary, synthetic oxytocin (Pitocin) is sometimes administered to facilitate uterine contractions.

Lactation and Maternal Bonding

Oxytocin is also crucial for lactation. When a baby begins to suckle at the mother's breast, sensory signals are transmitted to the brain, stimulating the release of oxytocin. This, in turn, triggers the contraction of cells around the milk-producing glands, leading to the ejection of milk—a process known as the "let-down reflex." Beyond its physiological role in breastfeeding, oxytocin fosters the emotional bond between mother and child, often enhancing the feelings of love and attachment.

Social Bonding and Emotional Regulation

Oxytocin's influence extends far beyond reproduction. Research has revealed that it plays a significant role in human relationships, particularly in the formation of social bonds and the regulation of emotions. One of the most well-documented effects of oxytocin is its ability to promote trust and empathy between individuals. This is particularly evident in romantic relationships, parent-child interactions, and close friendships.

The release of oxytocin can reduce stress, lower blood pressure, and promote feelings of well-being, often earning it the title of the "cuddle hormone." Studies have shown that oxytocin levels increase when people engage in positive social interactions, such as hugging, kissing, or simply being in the presence of loved ones. It has even been suggested that oxytocin may play a role in the development of social cognition—the ability to understand and interpret the emotions and intentions of others.

Oxytocin's effect on emotional regulation is also critical. In stressful situations, oxytocin can reduce anxiety and promote a calming effect. This has made it a target of interest in psychiatric research, as it may help alleviate symptoms in conditions such as social anxiety, PTSD, and depression. Additionally, oxytocin has been shown to play a role in enhancing positive emotional memories, further solidifying its position as a key hormone for emotional bonding and psychological well-being.

Pain Modulation and Immune Function

Interestingly, oxytocin also contributes to pain modulation. Studies have shown that oxytocin can reduce the perception of pain, acting as a natural analgesic. This has sparked interest in its potential therapeutic uses for chronic pain management and post-operative recovery. Furthermore, oxytocin has been implicated in modulating immune responses, promoting healing, and reducing inflammation.

The Endocrine and Psychosocial Interactions of Oxytocin

Beyond its individual physiological effects, oxytocin is integral to the interaction between the endocrine system and the brain, influencing various feedback loops within the body. The hormone is involved in regulating stress responses by interacting with cortisol and other stress hormones. It has been shown to counteract the negative effects of stress, including anxiety and depression, by promoting a sense of security and social support.

Moreover, the psychosocial aspect of oxytocin's influence is not to be underestimated. Studies have shown that people with higher oxytocin levels tend to have stronger social connections, better emotional health, and improved resilience to stress. This underscores the importance of oxytocin not only as a physiological regulator but also as a key player in emotional and social well-being.

Conclusion

Oxytocin is a multi-faceted hormone that plays an essential role in human physiology, affecting everything from childbirth to emotional regulation. Its far-reaching influence on both the body and the brain has made it an area of intense research and clinical interest. The discovery of oxytocin's diverse functions—ranging from its role in social bonding to its potential therapeutic uses in pain management, mental health, and beyond—has opened the door to new possibilities in medicine and psychology. As we explore the potential of oxytocin agonists in the following chapters, we will uncover how this powerful hormone can be harnessed to benefit human health, emotional well-being, and even social dynamics.

Chapter 2: What Are Oxytocin Agonists?

Definition and Function of Agonists

1. **Receptor Activation**

   Oxytocin agonists activate oxytocin receptors in the same way that natural oxytocin does. This can occur through a direct binding mechanism, where the agonist closely mimics the shape and chemical structure of oxytocin, enabling it to "fit" into the receptor binding site. Once attached, the agonist receptor complex stimulates downstream signaling pathways, such as the activation of phospholipase C (PLC), which generates inositol trisphosphate (IP3) and diacylglycerol (DAG). These second messengers increase the intracellular calcium levels, leading to cellular responses like muscle contraction or the release of neurotransmitters.

2. **Enhancement of Natural Effects**

   In some cases, oxytocin agonists can enhance the activity of endogenous oxytocin by increasing receptor sensitivity or prolonging the duration of receptor activation. For example, certain agonists can remain bound to the receptor longer than oxytocin itself, thus prolonging the physiological effect, such as the continuation of uterine contractions during labor. This property makes agonists useful in controlled medical applications, such as inducing labor or controlling postpartum hemorrhage.

3. **Selective Targeting**

Some oxytocin agonists are designed to selectively target specific receptor subtypes. The oxytocin receptor is widely expressed across the body, but its distribution varies in different tissues. For example, receptors in the uterus and brain may have slightly different structural conformations that can be selectively targeted by agonists. This specificity allows for a more refined approach in therapy, where oxytocin agonists can produce the desired physiological effects without overstimulating unintended systems.

Distinguishing Oxytocin Agonists from Antagonists

Understanding the difference between agonists and antagonists is critical to grasp the therapeutic potential of oxytocin-based treatments. While agonists activate the oxytocin receptor, antagonists block the receptor, preventing oxytocin from binding and exerting its natural effects.

Oxytocin antagonists, such as atosiban, are used in clinical settings to prevent premature labor by inhibiting uterine contractions. These compounds have the opposite effect of oxytocin agonists and are particularly useful in situations where preventing premature childbirth is necessary.

To fully appreciate the therapeutic potential of oxytocin agonists, it's essential to first understand the concept of an "agonist." In pharmacology, an agonist is a substance that binds to a specific receptor in the body and stimulates a biological response. In contrast to antagonists, which block or inhibit the receptor's activity, agonists activate receptors to induce a physiological effect.

Oxytocin agonists, therefore, are compounds that mimic or enhance the action of the natural hormone oxytocin by binding to its receptors. By interacting with the oxytocin receptor sites in the brain and peripheral tissues, these substances can produce effects similar to those induced by endogenous oxytocin—ranging from uterine contractions to social bonding and emotional regulation.

The concept of an agonist extends beyond mere mimicry. Oxytocin agonists can be designed to have a greater potency, a longer duration of action, or a more targeted effect compared to the body's own oxytocin. This ability to influence and enhance oxytocin activity has opened new avenues for therapeutic interventions, especially in fields like reproductive medicine, psychiatry, and pain management.

Mechanisms by Which Oxytocin Agonists Mimic or Enhance Oxytocin Activity

The mechanism of action for oxytocin agonists is rooted in their ability to bind to the oxytocin receptors (OTRs), which are G-protein coupled receptors (GPCRs) located in various tissues throughout the body. When oxytocin or its agonists bind to these receptors, a cascade of intracellular events is triggered that leads to the hormone's physiological effects.

The functional distinction between agonists and antagonists has far-reaching clinical implications. Agonists can be harnessed to enhance oxytocin's natural actions, such as promoting labor, improving emotional bonding, or reducing stress. Antagonists, on the other hand, can be used to counteract the hormone's effects, especially when the body's natural oxytocin levels are causing undesired responses, such as excessive uterine contractions.

Clinical and Therapeutic Applications of Oxytocin Agonists

The ability to manipulate oxytocin activity through its agonists has significant implications in both medical and psychological contexts. Below are some of the primary ways in which oxytocin agonists are being used:

1. Reproductive Medicine

In obstetrics, synthetic oxytocin (Pitocin) is commonly used as an agonist to induce labor, particularly in cases of prolonged labor or when the risk of complications necessitates delivery. By mimicking the natural actions of oxytocin, Pitocin stimulates uterine contractions and facilitates the progression of labor. Postpartum, oxytocin agonists are also used to control bleeding by promoting uterine contraction and reducing hemorrhage after childbirth.

2. Emotional and Social Bonding

Beyond its physiological role in reproduction, oxytocin agonists are being explored for their impact on emotional bonding. These compounds are particularly of interest in therapies aimed at enhancing emotional connections, improving social bonding, and increasing empathy. Studies have shown that oxytocin agonists can strengthen parent-child attachments, improve romantic relationships, and even foster trust in social interactions. The therapeutic potential in treating conditions like social anxiety, autism spectrum disorder (ASD), and even certain forms of depression has led to growing interest in oxytocin agonist research in psychiatry and psychology.

3. Pain Management

As a natural pain modulator, oxytocin's analgesic effects are being harnessed in clinical settings. Oxytocin agonists have been found to reduce the perception of pain, making them an attractive option for managing both acute and chronic pain. For example, they are being studied for use in post-operative care and in chronic pain conditions such as fibromyalgia and irritable bowel syndrome. Early studies suggest that oxytocin agonists could serve as adjuncts to other pain management therapies, potentially reducing the reliance on opioids and other painkillers.

4. Neuropsychological and Psychiatric Disorders

Oxytocin agonists are being investigated for their potential to treat various psychiatric disorders. In particular, disorders marked by social and emotional dysregulation, such as autism, social anxiety, and PTSD, are prime targets for oxytocin therapy. By enhancing oxytocin activity, these agonists could improve emotional regulation, reduce anxiety, and promote positive social interactions. Researchers are also exploring their potential in treating conditions such as depression, where emotional and social disconnection plays a key role.

5. Weight Management

Another intriguing area of research for oxytocin agonists is their role in appetite regulation and weight management. Studies have shown that oxytocin can influence feelings of satiety, helping to regulate food intake and maintain energy balance. Agonists that mimic or enhance oxytocin's actions in the hypothalamus may be used to treat obesity and metabolic disorders, offering a promising alternative to more traditional weight-loss interventions.

Conclusion

Oxytocin agonists represent a powerful tool in modern medicine, offering a broad range of therapeutic applications. By mimicking or enhancing the natural effects of oxytocin, these compounds can be used to induce labor, promote bonding, manage pain, and address a variety of psychiatric conditions. Their ability to activate oxytocin receptors opens up new frontiers in both clinical and psychological health, offering novel treatments for conditions that affect millions of people worldwide.

As research into oxytocin agonists continues, their potential will only expand. In the next chapters, we will delve into the various types of oxytocin agonists available today, their specific clinical applications, and the future of these compounds in improving human health and well-being.

Chapter 3: Types of Oxytocin Agonists

Oxytocin agonists come in various forms, each with unique properties and clinical applications. Understanding the differences between synthetic, natural, and emerging oxytocin agonists is essential for grasping their therapeutic potential and practical uses. This chapter will explore the key types of oxytocin agonists, including their origins, mechanisms, and applications in medicine.

1. Synthetic Oxytocin: Pitocin

Pitocin is the most widely known synthetic oxytocin agonist. It is a synthetic version of the naturally occurring hormone, oxytocin, and is used predominantly in reproductive medicine. Pitocin is identical to human oxytocin in its molecular structure and has the same biological effects, primarily stimulating uterine contractions.

Uses and Applications:

- **Labor Induction**: Pitocin is often used in hospital settings to induce or augment labor in pregnant women. If labor is slow, or if the health of the mother or fetus is at risk, Pitocin can help stimulate stronger, more frequent contractions to facilitate childbirth.

- **Postpartum Bleeding Control**: After childbirth, Pitocin is also used to control postpartum hemorrhage. By promoting uterine contraction, it helps to reduce bleeding and encourage the uterus to return to its normal size.

- **Post-Surgical Uterine Contraction**: In cases where a woman has had a cesarean section or other gynecological surgeries, Pitocin may be used to help the uterus contract and reduce blood loss.

Pitocin's major advantage lies in its predictable and controlled response. As it is synthesized in the laboratory, the concentration and dosage can be precisely managed, making it a reliable tool for labor management. However, its use is not without risks, including uterine hyperstimulation, which can lead to fetal distress or uterine rupture.

2. Natural and Semi–Synthetic Oxytocin Derivatives

In addition to Pitocin, there are natural and semi-synthetic derivatives of oxytocin that offer potential advantages in terms of bioactivity and side-effect profiles. These derivatives may be more readily absorbed by the body or possess longer half-lives, making them useful in different clinical contexts.

Natural Oxytocin:

Natural oxytocin, derived from animal sources, is less commonly used in modern clinical practice. Historically, oxytocin was extracted from the pituitary glands of cows and pigs for use in human medicine. This form of oxytocin is biochemically identical to human oxytocin, and its use was widespread prior to the development of synthetic versions like Pitocin.

Natural oxytocin's primary use was in managing labor and promoting uterine contractions during childbirth. However, due to concerns about animal sourcing and the availability of synthetic alternatives, the use of natural oxytocin has largely been phased out in favor of human-derived synthetics like Pitocin.

Semi–Synthetic Oxytocin Derivatives:

Semi-synthetic derivatives of oxytocin have been developed to provide better pharmacokinetics or to enhance specific therapeutic effects. These derivatives are chemically modified versions of natural or synthetic oxytocin, designed to have longer half-lives or altered receptor affinities. Some semi-synthetic oxytocin agonists are in the research phase and hold potential for more targeted and sustained therapeutic effects.

For example, **carbetocin**, a long-acting synthetic analog of oxytocin, has been developed for clinical use. Carbetocin is typically used for the prevention of postpartum hemorrhage following cesarean sections. It offers the advantage of longer duration compared to Pitocin, reducing the need for continuous infusion. Additionally, it has been shown to have fewer side effects, such as lower risks of uterine hyperstimulation, compared to Pitocin.

3. Emerging Compounds in the Research Phase

In recent years, the focus of oxytocin agonist research has expanded beyond synthetic and semi-synthetic derivatives of oxytocin. Researchers are developing novel oxytocin-based compounds that could offer new therapeutic applications, such as treating psychiatric disorders, managing chronic pain, or enhancing social functioning.

Oxytocin Agonists for Mental Health:

Emerging research has shown that oxytocin can have significant effects on emotional regulation, trust, and social behavior. This has led to the development of oxytocin agonists aimed at treating mental health disorders such as depression, social anxiety, and autism spectrum disorder (ASD).

A promising compound, **atosiban**, is a potent oxytocin receptor antagonist in its traditional use for inhibiting premature labor. However, modified forms of oxytocin agonists, such as those aimed at improving receptor specificity and brain penetration, are being developed to treat social deficits and emotional dysregulation seen in psychiatric conditions. These experimental oxytocin analogs are designed to cross the blood-brain barrier more effectively and target specific brain regions involved in emotional and social processing.

Oxytocin Agonists for Pain and Healing:

Oxytocin is a natural analgesic, with evidence showing that it can reduce pain perception and promote healing. Researchers are investigating the potential for new oxytocin agonists to be used as part of pain management protocols, particularly in chronic conditions such as fibromyalgia, irritable bowel syndrome, and post-operative recovery.

One area of research is developing **oxytocin analogs** that have selective effects on pain pathways. These compounds aim to enhance the analgesic properties of oxytocin while minimizing side effects like uterine contractions, which are undesirable in non-reproductive contexts.

The Search for More Selective Agonists:

To maximize the therapeutic potential of oxytocin, researchers are working on creating more selective and specialized agonists that activate specific oxytocin receptor subtypes. Different receptor subtypes are thought to mediate different effects—some that promote bonding and emotional connection, and others that modulate pain or stress. By targeting specific receptor populations, these newer compounds could provide more tailored treatments for a range of conditions without producing unwanted side effects.

For example, researchers are investigating **intranasal oxytocin agonists** that could be used to treat conditions like social anxiety or post-traumatic stress disorder (PTSD). Intranasal delivery allows the compound to bypass the blood-brain barrier more effectively, leading to faster and more targeted effects on brain function.

4. Delivery Systems for Oxytocin Agonists

The method of administration can have a significant impact on the efficacy and side-effect profile of oxytocin agonists. While Pitocin is typically administered intravenously in a hospital setting for labor induction, other forms of oxytocin agonists are being explored for different delivery methods, including:

- **Intranasal Sprays**: Intranasal oxytocin has gained popularity in clinical research for treating social and emotional disorders. This delivery method allows oxytocin to directly reach the brain, enhancing its effects on emotional regulation, trust, and social behavior. It is being studied for conditions such as autism spectrum disorder (ASD), social anxiety, and schizophrenia.

- **Oral and Sublingual Forms**: Oral administration of oxytocin is more challenging due to the hormone's degradation in the digestive system. However, advancements in drug formulation are paving the way for oral or sublingual oxytocin agonists that can achieve sufficient bioavailability to be effective for treating anxiety, depression, or other psychological conditions.

- **Injectables**: For more targeted and potent effects, injectable oxytocin agonists (similar to Pitocin) may be used in clinical settings for reproductive medicine or pain management. These formulations could be tailored to provide longer-lasting effects or reduce side effects associated with traditional oxytocin treatments.

Conclusion

Oxytocin agonists are an exciting and rapidly advancing area of medical science. From the widely used synthetic Pitocin to emerging compounds with selective receptor activity, these agonists offer diverse therapeutic potential for a range of medical and psychological conditions. As our understanding of oxytocin's role in human physiology expands, so too does the opportunity to develop more targeted, effective treatments.

The next chapters will explore specific applications of these oxytocin agonists in reproductive medicine, psychiatry, pain management, and more, showcasing their growing role in improving health and well-being.

Chapter 4: Oxytocin Agonists in Reproductive Medicine

Oxytocin agonists have become a cornerstone of reproductive medicine, particularly in managing childbirth and postpartum care. These compounds play crucial roles in regulating uterine contractions, controlling bleeding after delivery, and assisting in fertility treatments. In this chapter, we will explore the various applications of oxytocin agonists in reproductive medicine, highlighting their therapeutic benefits, safety concerns, and the ongoing debates around their use.

1. Inducing Labor and Controlling Postpartum Bleeding

One of the most common uses of oxytocin agonists, particularly synthetic oxytocin (Pitocin), is in **labor induction**. While labor generally starts naturally when the body signals the need to deliver, there are circumstances where it is medically necessary to induce or augment labor. For example:

- **Post-term Pregnancy**: If a pregnancy extends past the 42-week mark, there is a higher risk of complications such as decreased amniotic fluid, placenta insufficiency, or fetal distress. In such cases, Pitocin may be used to initiate contractions.
- **Medical Conditions**: Pregnancies complicated by conditions such as preeclampsia, gestational diabetes, or maternal infections might also require labor induction to prevent further complications for both mother and baby.
- **Fetal Distress**: If there are signs of fetal distress, such as abnormal heart rates, a controlled induction of labor can be used to ensure a safer delivery.

Pitocin for Inducing Labor:

Pitocin is administered via intravenous (IV) infusion to gradually stimulate uterine contractions. It mimics the natural action of oxytocin by binding to oxytocin receptors in the uterine muscles, resulting in rhythmic contractions that help to expel the baby.

While Pitocin is highly effective in initiating labor, it requires careful monitoring due to the risk of **uterine hyperstimulation**, which can lead to fetal distress or uterine rupture. This condition occurs when contractions become excessively frequent or intense, potentially compromising oxygen flow to the fetus.

Controlling Postpartum Hemorrhage:

After childbirth, one of the most dangerous complications is excessive bleeding, known as **postpartum hemorrhage** (PPH). Oxytocin agonists, including Pitocin and its derivatives, are often used to manage and prevent this life-threatening condition. By promoting continued uterine contractions after delivery, oxytocin helps constrict blood vessels in the uterus, reducing blood loss and promoting uterine involution (the process by which the uterus returns to its pre-pregnancy size).

- **Pitocin Infusion**: Following delivery, a continuous IV infusion of Pitocin is commonly used to ensure adequate uterine tone and prevent excessive bleeding.
- **Carbetocin**: A longer-acting oxytocin analog, **carbetocin** has been increasingly used in cesarean deliveries to prevent postpartum hemorrhage. It is preferred in certain cases because it has a longer duration of action compared to Pitocin, reducing the need for constant monitoring.

Despite its efficacy, the use of oxytocin agonists for controlling postpartum bleeding is not without its risks. Overuse or rapid infusion of oxytocin can result in **uterine rupture** or **water intoxication** (when excessive amounts of fluid are absorbed by the body, potentially leading to electrolyte imbalances). As such, careful dosing and patient monitoring are essential.

2. Applications in Fertility Treatments

Oxytocin's role extends beyond labor induction and postpartum care to **fertility treatments**. While oxytocin itself is not a fertility drug, its agonists are used in certain assisted reproductive techniques to improve outcomes, especially in **ovarian stimulation** and **embryo transfer** processes.

Enhancing Uterine Contractions:

In fertility treatments such as in vitro fertilization (IVF), oxytocin agonists may be used to assist with the **implantation of embryos**. For the embryo to successfully implant into the uterine lining, the uterus must be adequately prepared. Sometimes, oxytocin agonists are used to encourage mild uterine contractions that improve blood flow to the uterus, which can help the embryo adhere to the endometrial lining.

Timing of Embryo Transfer

carbetocin

Luteal Phase Support:

In IVF, the luteal phase (the second half of the menstrual cycle after ovulation) is crucial for embryo survival. Adequate hormone support, often through synthetic progesterone, is needed for the embryo to implant and thrive. Research has explored the possibility that **oxytocin agonists** could help support the luteal phase by promoting uterine contractility and enhancing progesterone's effects on the uterine lining.

Ovarian Hyperstimulation Syndrome (OHSS):

A potential complication of fertility treatments, particularly those involving **gonadotropins** to stimulate the ovaries, is **ovarian hyperstimulation syndrome** (OHSS), a condition where the ovaries become swollen and painful. Oxytocin agonists are not typically used to treat OHSS directly, but in some cases, they may be explored for their ability to regulate fluid balance and reduce ovarian swelling through their vasopressor effects.

3. Safety Concerns and Controversies

Despite their effectiveness in managing labor and fertility treatments, the use of oxytocin agonists is not without controversy and safety concerns. Understanding these issues is crucial for healthcare providers to balance the benefits of these treatments with the associated risks.

Uterine Hyperstimulation and Fetal Distress:

As previously mentioned, excessive stimulation of the uterus through oxytocin agonists can lead to uterine hyperstimulation, where contractions become too frequent and intense. This can reduce the oxygen supply to the fetus and increase the risk of **fetal distress**, which may necessitate emergency interventions, including cesarean delivery.

Management of Hyperstimulation

Water Intoxication and Hyponatremia:

Oxytocin is a peptide hormone, and excessive amounts can lead to **water intoxication**, which occurs when the body takes in too much water, diluting blood sodium levels. In severe cases, this condition can result in **hyponatremia**, which can lead to seizures, coma, and even death. Healthcare professionals must be vigilant about limiting the amount of fluid infused along with oxytocin.

Risk of Uterine Rupture:

Another risk associated with high doses of oxytocin is **uterine rupture**, a serious and life-threatening condition where the uterus tears, often during labor. This is a rare but dangerous complication that requires immediate medical attention and usually necessitates an emergency cesarean section.

Precautions in Induction

4. The Future of Oxytocin Agonists in Reproductive Medicine

While synthetic oxytocin and its derivatives have been in clinical use for decades, ongoing research is exploring new possibilities for improving their use in reproductive medicine.

- **Long-Acting Oxytocin Agonists**: New formulations of oxytocin agonists that have longer half-lives, like **carbetocin**, are already showing promise in reducing postpartum bleeding, especially following cesarean deliveries. The future may see more such agents developed to reduce the need for multiple dosing and continuous monitoring.

- **Targeted Therapies for Labor and Fertility**: As our understanding of the oxytocin system improves, there is potential for more **targeted therapies** that can specifically modulate uterine contractions without affecting other organs. This could reduce side effects like uterine rupture and improve the safety and efficacy of labor induction and fertility treatments.

The potential for oxytocin agonists in reproductive medicine is vast, and as research advances, new applications and safer methods of delivery will emerge, offering women more effective and less risky options for childbirth and fertility management.

Conclusion

Oxytocin agonists, particularly synthetic versions like Pitocin, have revolutionized reproductive medicine, enabling controlled labor induction, effective management of postpartum hemorrhage, and enhanced fertility treatment outcomes. However, their use must be carefully monitored due to the associated risks, including uterine hyperstimulation, water intoxication, and uterine rupture. As research progresses, the development of more targeted, long-acting oxytocin analogs holds great promise for improving maternal and fetal health in reproductive medicine. The continued exploration of these therapies, alongside advancements in safety protocols, will pave the way for more refined and effective interventions in the field.

Chapter 5: The Role of Oxytocin Agonists in Bonding and Attachment

Oxytocin is widely recognized for its profound impact on **human bonding and attachment**, influencing relationships between parents and children, romantic partners, and even individuals within broader social networks. Often referred to as the "love hormone," oxytocin plays a critical role in the formation and maintenance of deep, meaningful connections. This chapter explores how oxytocin agonists—synthetic or naturally derived compounds that enhance or mimic the action of oxytocin—can influence bonding, attachment, and emotional intimacy.

1. Oxytocin's Role in Maternal Bonding

Maternal bonding is a fundamental aspect of human development, influencing the long-term emotional health of both mother and child. The relationship between **oxytocin** and maternal bonding is well-established, with oxytocin levels rising during childbirth, lactation, and early interactions between mother and infant. This hormone enhances maternal behaviors, such as caregiving, nurturing, and protecting the infant.

Mechanisms of Maternal Bonding:

- **Labor and Delivery**: Oxytocin plays an essential role in the birth process by facilitating uterine contractions and promoting the delivery of the placenta. After childbirth, it continues to play a crucial role in helping the mother bond with the newborn. High levels of oxytocin are associated with increased maternal responsiveness, emotional attachment, and positive caregiving behaviors.

- **Breastfeeding**: When the baby suckles at the breast, oxytocin is released, which triggers **milk ejection** and strengthens the bond between mother and child. The act of breastfeeding itself further reinforces the emotional connection, promoting a sense of calm, security, and mutual trust between mother and infant.

In clinical settings, **oxytocin agonists** may be used to enhance maternal bonding, particularly in cases of preterm birth, difficult deliveries, or situations where early mother-infant contact is limited. For example, the administration of oxytocin may be used to promote lactation or encourage bonding in mothers who are struggling with initial breastfeeding.

Clinical Use:

Preterm Birth

2. Influence on Father–Child Bonding

While much of the literature on oxytocin and bonding focuses on the mother-child dyad, there is growing recognition of oxytocin's role in **father-child bonding** as well. Fathers also experience a rise in oxytocin levels during labor and delivery, and particularly in the early stages of fatherhood. This increase in oxytocin has been shown to enhance paternal behaviors and emotional responsiveness toward the newborn.

Evidence from Research:

- **Father-Infant Interaction**: Studies have shown that fathers who experience higher levels of oxytocin during the birth process or in the early postpartum period are more likely to engage in **positive caregiving behaviors**, such as holding, comforting, and communicating with their child. This fosters a secure attachment between father and child, which can have long-term effects on the child's emotional development.

- **Fathering and Stress**: Oxytocin's ability to reduce stress and anxiety in the parent is another critical aspect. It may act as a buffer against the stresses of new fatherhood, helping to regulate the father's emotions and promoting a calmer, more responsive interaction with the infant.

Therapeutic Use:

In situations where a father's bonding is delayed, such as in cases of **postpartum depression** or **post-traumatic stress disorder** (PTSD) following childbirth, oxytocin agonists could potentially be used to help facilitate emotional engagement. The administration of oxytocin might promote feelings of connection and emotional warmth, aiding the father in building a stronger relationship with the child during the crucial early stages.

3. Oxytocin Agonists in Relationship and Partner Bonding

While oxytocin is most commonly associated with the parent-child relationship, its role in **romantic and partner bonding** is equally significant. Oxytocin is released during intimate physical contact, such as hugging, kissing, and sexual intercourse. It is also believed to play a part in **trust, empathy, and emotional closeness** between romantic partners, fostering relationship satisfaction and intimacy.

Enhancing Emotional Intimacy:

Oxytocin's ability to promote **empathy** and reduce social anxiety makes it a key player in relationship dynamics. When oxytocin levels rise in both partners, there is often an increase in **emotional reciprocity**, wherein both individuals feel more emotionally attuned to one another. This can help foster a sense of security and mutual understanding in relationships.

In clinical settings, **oxytocin agonists** are sometimes used to help couples experiencing relationship difficulties, such as trust issues, communication problems, or emotional distance. By enhancing emotional connection, oxytocin-based therapies could support couples in re-establishing trust and intimacy.

Partner Bonding and Therapy:

Couples experiencing emotional or physical disconnection, particularly following trauma or infidelity, might benefit from therapies that promote oxytocin release. These might include **oxytocin agonists** administered in clinical settings, or through therapeutic touch and practices such as couples' massage, which naturally stimulates oxytocin production.

4. Clinical Uses in Enhancing Attachment in Adoption or Foster Care

For children placed in adoptive or foster homes, attachment disorders can present significant challenges. These children often face emotional difficulties related to past trauma, abuse, or neglect, making it difficult for them to form healthy bonds with their new caregivers.

Oxytocin and Attachment Disorders:

Oxytocin agonists hold promise as a potential treatment for enhancing **attachment in children** who have experienced early trauma. Research suggests that children who experience high levels of early stress or neglect may have lower levels of oxytocin, which could impair their ability to trust and connect with others.

In clinical trials, oxytocin has been tested as part of therapies for children with **attachment disorders**, helping to promote emotional bonding with adoptive or foster parents. Studies suggest that oxytocin may help these children develop **secure attachments**, encouraging trust and emotional engagement with their new caregivers.

Applications in Therapy:

- **Attachment-Based Therapy**: Oxytocin agonists could potentially be integrated into **attachment-based therapies** for children and families. By improving oxytocin signaling, these therapies could help children feel more secure and reduce symptoms of anxiety, depression, or aggression often seen in children with insecure attachments.
- **Trauma Recovery**: Children in foster or adoptive care often carry the scars of early trauma, which can make bonding difficult. Oxytocin-based interventions may help to regulate emotional responses, allowing for healthier emotional development and facilitating better attachment behaviors.

5. Social and Behavioral Implications

Oxytocin's impact on bonding and attachment extends beyond the family unit, influencing **broader social interactions** and group dynamics. Enhanced oxytocin activity can promote greater **social cohesion**, cooperation, and altruism, all of which contribute to the development of strong social bonds.

In Therapeutic Settings:

In addition to its effects on familial and romantic relationships, oxytocin's influence on attachment has implications for **psychotherapy** and **relationship counseling**. Therapists working with clients who struggle with interpersonal relationships, whether due to social anxiety, trauma, or past attachment issues, might use oxytocin agonists to facilitate deeper emotional connections and improve the effectiveness of therapy.

6. Safety, Ethics, and Considerations in Bonding Interventions

While oxytocin agonists hold promise for enhancing bonding and attachment, their use requires careful consideration. The ethical implications of manipulating oxytocin levels to influence emotional states and behaviors are significant, particularly when it comes to vulnerable populations such as children or individuals recovering from trauma.

- **Risks of Misuse**: There is a potential for the misuse of oxytocin to manipulate emotional responses artificially. For example, using oxytocin agonists to influence the emotional responses of children or patients could lead to unintended psychological consequences, such as emotional dependency or lack of autonomy.

- **Long-Term Effects**: The long-term effects of oxytocin agonists on emotional regulation and bonding are not yet fully understood. Ongoing research will be critical to ensuring that these interventions are safe and effective for enhancing attachment without causing negative consequences.

Conclusion

Oxytocin agonists offer valuable therapeutic potential in promoting bonding and attachment across a variety of contexts, from maternal and paternal bonding to fostering emotional connections in romantic relationships, and even in adoption and foster care settings. By enhancing the natural effects of oxytocin, these compounds can support emotional intimacy, reduce anxiety, and encourage secure attachments. However, as with all therapeutic interventions, the use of oxytocin agonists requires careful consideration of safety, ethical implications, and long-term effects. As research progresses, oxytocin agonists may play an increasingly important role in fostering healthier, more connected individuals and families.

Chapter 6: Emotional and Social Effects of Oxytocin Agonists

Oxytocin is renowned for its effects on human emotions and social behaviors, earning it the nickname "the **love hormone**." Its role in fostering trust, empathy, emotional connection, and social bonding has profound implications for both personal well-being and societal functioning. This chapter delves into the emotional and social effects of **oxytocin agonists**, examining how these compounds can enhance human interactions, promote positive social behaviors, and improve emotional regulation.

1. Enhancing Trust and Empathy

At the core of many human interactions is **trust**—the foundational element of social cohesion. Oxytocin is directly involved in the development of trust between individuals, playing a key role in establishing secure, positive relationships. Studies have shown that **oxytocin agonists** can enhance the sense of trust, particularly in situations where social bonds are initially weak or where trust has been damaged.

Trust in Social Interactions:

- **Studies on Trust**: Research has shown that individuals administered oxytocin agonists display an increased willingness to trust others, even in unfamiliar or ambiguous situations. Oxytocin enhances the **reliability of social signals** and reduces suspicion, making people more open and receptive in social exchanges.
- **Trust in Relationships**: In romantic relationships or friendships, oxytocin promotes emotional security by encouraging vulnerability. By enhancing trust, oxytocin agonists can improve communication, reduce conflicts, and foster deeper intimacy between partners.

Empathy and Emotional Understanding:

- **Empathy Enhancement**: Empathy—the ability to understand and share the feelings of others—is a key driver of social connection. Oxytocin has been shown to enhance **emotional empathy**, allowing individuals to better recognize and resonate with the emotional states of others. This makes oxytocin agonists particularly valuable in therapeutic settings where emotional understanding is central to the healing process.

- **Empathy in Therapy**: In clinical practices, such as **counseling** or **psychotherapy**, oxytocin agonists may be used to help individuals with emotional detachment or those struggling with **social difficulties** (e.g., social anxiety or autism spectrum disorder). By enhancing empathetic responses, these compounds can help individuals engage more meaningfully with others, improving therapeutic outcomes.

2. Strengthening Emotional Connection

Oxytocin is deeply intertwined with **emotional connection**, both between intimate partners and within broader social contexts. The emotional warmth and sense of belonging fostered by oxytocin can significantly impact social relationships, promoting behaviors that enhance both individual well-being and group harmony.

Bonding Beyond the Immediate Family:

- **Romantic Relationships**: Oxytocin plays a critical role in the bonding between romantic partners. During physical touch, intimacy, and shared positive experiences, oxytocin levels rise, enhancing **emotional intimacy** and reinforcing affectionate behaviors. Oxytocin agonists can potentially be used to **enhance relationship satisfaction**, particularly in individuals facing emotional barriers or emotional disconnect.

- **Group Cohesion**: Oxytocin also promotes social cohesion in larger groups. It encourages **cooperative behavior**, emotional regulation, and conflict resolution, making it an essential component of effective teamwork, community-building, and organizational success. By enhancing oxytocin activity, individuals may experience a greater sense of **connectedness** and shared purpose within groups or societies.

Applications in Therapy:

In couples therapy, oxytocin agonists can be used to help partners reconnect emotionally. In cases where **emotional withdrawal** has occurred—due to stress, infidelity, or long-term relationship issues—oxytocin agonists might be employed to help restore emotional closeness, fostering an environment of trust and understanding.

For individuals undergoing **trauma recovery**, oxytocin agonists may help to re-establish a sense of emotional security and help people feel safer in emotional interactions, leading to more resilient relationships.

3. Oxytocin Agonists and Social Behavior

Social behavior is driven by a complex interplay of emotions, cognition, and environmental factors. Oxytocin is integral to shaping social behavior, facilitating positive interactions, and regulating social anxiety. The emotional warmth induced by oxytocin influences how individuals relate to one another, impacting everything from casual conversations to complex social dynamics.

Social Behavior in Groups:

- **Social Affiliation**: Oxytocin agonists can enhance the **sense of social affiliation**, promoting more prosocial behaviors. In group settings, individuals under the influence of oxytocin are more likely to engage in cooperative actions, share resources, and exhibit **altruistic tendencies**. These behaviors promote **social harmony** and contribute to the development of **stronger social networks**.

- **Reduced Prejudice**: Some studies have suggested that oxytocin may also reduce **prejudices and stereotypes**. By enhancing empathy and emotional connection, oxytocin agonists can promote more inclusive and compassionate social attitudes, which could be valuable in addressing social inequality and discrimination.

Therapeutic Applications:

In mental health treatment, oxytocin agonists can assist individuals with **social anxiety** or those who experience emotional difficulties in group settings. By boosting oxytocin levels, individuals may feel more comfortable engaging with others and more at ease in social situations. **Cognitive-behavioral therapy (CBT)**, often used for social anxiety, may be enhanced by oxytocin, leading to improved outcomes in social skill training and emotional regulation.

4. Reducing Stress and Promoting Emotional Regulation

Oxytocin's calming effects are essential to its role in emotional regulation. Oxytocin is often referred to as a **stress-buffering hormone**, as it has the ability to reduce **cortisol** levels, a primary stress hormone. By reducing stress, oxytocin promotes emotional stability, helping individuals manage both acute and chronic stress more effectively.

The Stress-Reduction Mechanism:

- **Cortisol Regulation**: Oxytocin works in conjunction with the autonomic nervous system to reduce the **physiological symptoms of stress**—such as elevated heart rate, blood pressure, and anxiety. When oxytocin agonists are administered, the body is better able to return to a state of equilibrium after stressful events.

- **Emotional Resilience**: The calming effects of oxytocin contribute to **emotional resilience**, particularly in individuals who face chronic stress or trauma. By promoting a **calm state of mind**, oxytocin enables individuals to process emotional stimuli more effectively, improving their ability to respond to challenging or distressing situations with composure.

Applications in Stress-Related Disorders:

For individuals suffering from **post-traumatic stress disorder (PTSD)**, **generalized anxiety disorder (GAD)**, or **social anxiety**, oxytocin agonists can play an important role in **alleviating the emotional dysregulation** commonly associated with these conditions. Studies suggest that oxytocin's ability to enhance social support networks and reduce anxiety symptoms could make it a promising addition to therapeutic approaches for these disorders.

5. Social Implications for Therapy and Relationship Counseling

The use of oxytocin agonists in therapy has far-reaching implications, particularly in the fields of **relationship counseling** and **psychiatric treatment**. By enhancing emotional connection and empathy, oxytocin agonists can potentially improve the therapeutic relationship, facilitating **greater openness and trust** between therapist and client. This can lead to **more effective treatment outcomes**, especially in individuals with attachment disorders, emotional trauma, or social difficulties.

Relationship Counseling:

- **Healing from Trauma**: For individuals experiencing relationship difficulties due to past trauma, oxytocin agonists may assist in fostering emotional healing and **relationship reconnection**. By reducing emotional distance and promoting empathy, oxytocin can be a vital component of **couples' therapy** or **family therapy**.

- **Improving Communication**: In couples therapy, oxytocin agonists might be used to enhance **communication skills**, allowing partners to express their feelings more freely and to better understand each other's emotional needs. This can lead to more positive relationship dynamics and deeper emotional intimacy.

6. Ethical Considerations and Potential Risks

While oxytocin agonists hold great promise for enhancing emotional and social outcomes, their use must be carefully considered from an **ethical perspective**. The ability to influence emotions, behaviors, and relationships raises important questions about consent, manipulation, and the potential for misuse.

Risks of Manipulation:

- The ethical dilemma of using oxytocin agonists to influence **emotional states** or **social behavior** is significant. While these compounds can be beneficial for promoting emotional connection, their use in manipulating social dynamics or altering emotional responses without full consent could have **unintended consequences**.
- Long-term use or over-reliance on oxytocin agonists could also potentially lead to **emotional dependency**, where individuals might rely on external compounds to maintain emotional balance, rather than developing their own coping mechanisms.

Conclusion

Oxytocin agonists offer remarkable potential for enhancing emotional connection, social behavior, and emotional regulation. By fostering trust, empathy, and cooperative behavior, these compounds can improve personal relationships, support social cohesion, and alleviate the symptoms of stress and anxiety. In therapeutic settings, oxytocin agonists provide a unique tool for enhancing emotional understanding and connection, while offering new avenues for treating disorders linked to social and emotional dysfunction. However, as with all therapeutic interventions, careful consideration must be given to the ethical implications of their use and the long-term effects on emotional well-being. As research progresses, oxytocin agonists could play an increasingly pivotal role in shaping the future of social and emotional health.

Chapter 7: Oxytocin and the Brain

The neurobiological effects of **oxytocin** on the brain are profound and far-reaching. As a hormone and neurotransmitter, oxytocin influences both physiological and psychological processes, particularly those related to social behaviors, emotional regulation, and cognitive function. This chapter explores the intricate ways in which oxytocin interacts with the brain, its role in modulating mood, anxiety, and stress, and its potential therapeutic applications in neuropsychological disorders.

1. Neurobiology of Oxytocin Receptors in the Brain

To understand how oxytocin affects the brain, it's essential to first look at its receptors. Oxytocin receptors are distributed throughout various brain regions, including areas involved in emotion regulation, social cognition, and memory. These receptors are part of the **G-protein coupled receptor family**, and their activation triggers a cascade of intracellular events that modulate brain function.

Key Brain Regions Affected by Oxytocin:

- **Amygdala**: The amygdala plays a central role in emotional processing, particularly in detecting threats and regulating fear responses. Oxytocin receptors in the amygdala are involved in **reducing fear and anxiety** while promoting **social approach behaviors**. The release of oxytocin in this region helps mitigate overactive responses to stressors, supporting emotional regulation and social interactions.

- **Prefrontal Cortex**: The prefrontal cortex is responsible for higher cognitive functions, such as decision-making, impulse control, and emotional regulation. Oxytocin's action on this area enhances **empathy, moral decision-making**, and the ability to engage in **prosocial behaviors**. It also strengthens **social cognition**, which refers to the ability to interpret and respond to social cues.

- **Hippocampus**: The hippocampus, which is crucial for memory and learning, also contains oxytocin receptors. Oxytocin's effects on the hippocampus help modulate emotional memories, enabling individuals to integrate past experiences with current emotional responses in a way that supports **positive social bonding**.

- **Ventral Striatum**: This region is associated with the reward system and the experience of pleasure. Oxytocin's influence here suggests that it may enhance **reward processing** during social interactions, further reinforcing prosocial behavior and attachment formation.

2. Oxytocin's Impact on Mood, Anxiety, and Stress

Oxytocin is often associated with positive emotional states, particularly feelings of connection, trust, and safety. Its effects on mood regulation, anxiety reduction, and stress management make it an important player in emotional health and well-being.

Mood Regulation:

- **Oxytocin and Positive Emotion**: In both animals and humans, elevated oxytocin levels are linked to feelings of **contentment, happiness, and emotional security**. The hormone promotes a sense of calm and well-being, which is critical for maintaining **emotional balance** during stressful situations.

- **Oxytocin and Depression**: Some research suggests that oxytocin might play a role in alleviating symptoms of **depression**, particularly in individuals who experience depression linked to social isolation or poor social bonding. Oxytocin's ability to enhance emotional connection and empathy might help individuals feel less isolated and more emotionally supported.

Anxiety and Stress Reduction:

- **Oxytocin as a Stress Buffer**: Oxytocin is considered a **stress-buffering hormone**, working in concert with other neuropeptides like **cortisol** to reduce the physiological and psychological effects of stress. By binding to receptors in the **hypothalamus** and **pituitary gland**, oxytocin decreases the production of **stress hormones**, leading to lower blood pressure, heart rate, and reduced anxiety.

- **Effect on Social Anxiety**: Studies have shown that oxytocin can reduce social anxiety, making it easier for individuals to navigate social interactions without excessive fear of judgment or rejection. This effect is especially helpful for people with **social anxiety disorder (SAD)** or **generalized anxiety disorder (GAD)**, where fear of social situations is a prominent symptom.

3. Therapeutic Potential in Neuropsychological Disorders

Oxytocin's role in modulating emotional states and promoting social behaviors suggests that it may have therapeutic potential in a variety of **neuropsychological disorders**, particularly those that involve emotional dysregulation, social difficulties, and impaired bonding.

Oxytocin and Autism Spectrum Disorder (ASD):

- **Social Cognition and ASD**: One of the hallmark features of autism spectrum disorder is difficulty with **social cognition**—the ability to interpret and respond to social cues. Oxytocin has shown promise in enhancing social cognition and improving interpersonal interactions in individuals with ASD. Clinical trials have suggested that oxytocin administration can improve **eye contact, social engagement**, and emotional reciprocity.

- **Improved Social Functioning**: Research indicates that oxytocin may help individuals with ASD build better relationships with family members, peers, and caregivers. By enhancing the **empathy** and **emotional responsiveness** necessary for social bonding, oxytocin could play a role in fostering more meaningful social connections for those with autism.

Oxytocin in Schizophrenia:

- **Emotional Regulation and Psychosis**: In individuals with **schizophrenia**, difficulties with emotional regulation and social cognition often exacerbate symptoms. Oxytocin's potential to **modulate emotional responses** and enhance **social perception** could provide a therapeutic advantage for these individuals. Some studies have shown that oxytocin administration can improve **symptoms of psychosis**, particularly those related to **social withdrawal** and **paranoia**.

- **Improving Social Functioning**: By boosting oxytocin levels, patients with schizophrenia may experience improved social interactions, leading to better integration in communities and enhanced quality of life. This may also have a secondary benefit of improving adherence to treatment plans and reducing the risk of relapse.

Oxytocin and Post–Traumatic Stress Disorder (PTSD):

- **Trauma and Attachment**: PTSD is marked by heightened emotional responses to traumatic memories, social withdrawal, and difficulty in forming secure attachments. Oxytocin's ability to foster **secure attachment bonds** could make it a promising candidate for treating PTSD. By enhancing **trust** and **emotional connection**, oxytocin could help individuals rebuild healthy relationships and reduce hyperarousal symptoms associated with trauma.

- **Reduced Hyperarousal**: In clinical settings, oxytocin has shown the potential to reduce symptoms of **hyperarousal** (such as increased heart rate, irritability, and heightened startle response) that often accompany PTSD. Additionally, oxytocin's calming effects can help individuals process traumatic memories with less emotional distress.

4. The Role of Oxytocin in Memory and Learning

Beyond its emotional and social functions, oxytocin also plays a role in memory formation and learning. By acting on the **hippocampus**, oxytocin can influence how emotional memories are encoded and recalled, potentially improving the emotional context of past experiences.

Enhancing Memory Consolidation:

- **Emotional Memories**: Research has shown that oxytocin enhances the **consolidation of emotional memories**, particularly those related to social interactions. This means that positive emotional experiences and social bonds are more likely to be remembered and recalled, which may be beneficial in treating conditions such as **depression** or **post-traumatic stress disorder (PTSD)**, where individuals often have trouble recalling positive experiences.

- **Cognitive Functioning**: Preliminary research suggests that oxytocin might also influence cognitive functions such as **attention** and **learning ability**. In people with cognitive impairments or neurodegenerative diseases (such as Alzheimer's), oxytocin's effects on memory might provide a novel avenue for treatment.

5. Future Directions in Oxytocin Research

As our understanding of oxytocin's neurobiological effects grows, researchers are beginning to explore its therapeutic potential in a wide range of **neuropsychological disorders**. Future studies will need to address several key areas to optimize the clinical use of oxytocin agonists:

- **Optimal Dosage and Delivery Methods**: Determining the right **dosage**, timing, and delivery methods for oxytocin is essential for maximizing its therapeutic benefits while minimizing side effects. Whether delivered through nasal sprays, injections, or oral formulations, the most effective method of administration for various disorders remains an area of active investigation.

- **Personalized Treatment Protocols**: As with many modern treatments, **personalized medicine** may play a significant role in determining the appropriate use of oxytocin agonists. Individual differences in **genetic makeup** and **neurobiology** will likely influence how people respond to oxytocin, and understanding these variations will help tailor treatments to maximize effectiveness.

Conclusion

Oxytocin's impact on the brain is vast and multifaceted, influencing everything from emotional regulation and social bonding to memory and cognitive function. By acting on key brain regions like the **amygdala**, **prefrontal cortex**, and **hippocampus**, oxytocin plays a critical role in shaping both emotional experiences and social interactions. Its therapeutic potential in neuropsychological disorders—such as **autism spectrum disorder (ASD)**, **schizophrenia**, and **post-traumatic stress disorder (PTSD)**—holds great promise for improving the quality of life for individuals with these conditions. As research advances, oxytocin may become an increasingly important tool in addressing emotional and cognitive dysfunctions, offering new hope for those struggling with mental health challenges.

Chapter 8: Oxytocin Agonists and Anxiety Disorders

Anxiety disorders are among the most prevalent mental health conditions worldwide, affecting millions of people across all demographics. From generalized anxiety disorder (GAD) and social anxiety to more specific forms such as post-traumatic stress disorder (PTSD), these conditions often share a core feature: the disruption of emotional regulation, particularly in stressful or socially charged situations. Given oxytocin's key role in modulating emotional responses, its agonists have emerged as promising therapeutic agents for alleviating anxiety symptoms. This chapter explores the mechanisms by which oxytocin agonists reduce anxiety, their potential use in treating anxiety disorders, and the clinical studies that provide insight into their efficacy.

1. The Mechanisms of Action: How Oxytocin Agonists Influence Anxiety

Oxytocin's influence on anxiety is primarily rooted in its ability to affect the brain's **emotional processing circuits**, particularly areas that govern fear, social bonding, and stress regulation. The hormone interacts with several key brain regions, including the **amygdala**, **prefrontal cortex**, and **hippocampus**, which are crucial for processing emotions and managing anxiety.

Modulating the Amygdala:

The **amygdala**, a small almond-shaped cluster of nuclei located in the temporal lobe, plays a central role in the brain's fear and anxiety responses. Oxytocin reduces hyperactivity in the amygdala, effectively **dampening fear responses** to perceived threats. By doing so, oxytocin agonists can help mitigate the exaggerated fear response seen in conditions like **social anxiety** and **PTSD**, where individuals often react disproportionately to social stimuli or traumatic memories.

Enhancing Prefrontal Cortex Regulation:

The **prefrontal cortex** (PFC) is responsible for higher-order functions, including **emotion regulation**, decision-making, and **impulse control**. Oxytocin's action on the PFC can help improve **cognitive reappraisal**—the ability to reinterpret stressful or negative situations in a less threatening light. By improving the PFC's ability to regulate emotions, oxytocin agonists may enhance an individual's resilience to anxiety-inducing situations, reducing the intensity of emotional responses and promoting a more balanced perspective.

Strengthening Social Bonding:

Social connectedness and support are crucial for managing anxiety. Oxytocin is often referred to as the "bonding hormone" because of its pivotal role in promoting trust, empathy, and emotional connection. Oxytocin agonists enhance social bonding, which can act as a natural buffer against anxiety, especially in social contexts. For individuals with **social anxiety disorder (SAD)**, where fear of social judgment is prevalent, the administration of oxytocin can **reduce social avoidance** and promote greater social engagement, helping individuals face social situations with more confidence.

2. Oxytocin Agonists in Treating Specific Anxiety Disorders

The therapeutic potential of oxytocin agonists has been investigated in a variety of anxiety disorders, including **generalized anxiety disorder (GAD)**, **social anxiety disorder (SAD)**, and **post-traumatic stress disorder (PTSD)**. Below, we explore how oxytocin's anxiolytic (anxiety-reducing) effects may provide relief for these conditions.

Generalized Anxiety Disorder (GAD):

GAD is characterized by chronic and excessive worry about everyday events, often accompanied by physical symptoms such as restlessness, muscle tension, and sleep disturbances. Individuals with GAD frequently experience heightened **sensory sensitivity** and **hypervigilance**, which can make even normal stressors feel overwhelming. Oxytocin agonists have shown potential in reducing these symptoms by promoting relaxation and improving **emotional regulation**. Some studies suggest that oxytocin administration can reduce **cortisol levels**, the body's primary stress hormone, which may help individuals with GAD better cope with anxiety-provoking situations.

Social Anxiety Disorder (SAD):

Social anxiety disorder is marked by an intense fear of social situations, often accompanied by the fear of being judged or negatively evaluated by others. Social interactions become daunting and may even be avoided altogether. Clinical research has shown that oxytocin agonists can reduce anxiety in social settings by **increasing trust** and **decreasing fear** of negative evaluation. One notable study demonstrated that **nasal oxytocin** administration helped individuals with SAD engage more readily in social conversations, improved their self-confidence, and reduced their physiological anxiety response (e.g., heart rate, sweating). This effect appears to be linked to oxytocin's ability to **enhance emotional empathy** and make social interactions feel safer and less threatening.

Post-Traumatic Stress Disorder (PTSD):

PTSD is a severe anxiety disorder that can develop after exposure to traumatic events, leading to **flashbacks, nightmares**, and a pervasive sense of fear and helplessness. Individuals with PTSD often struggle with emotional numbing, avoidance behaviors, and hyperarousal, which can interfere with their ability to process trauma and form secure relationships. Oxytocin has been shown to play a role in **modulating fear memories** and **promoting emotional regulation**, which may help individuals with PTSD process traumatic memories in a safer, less distressing manner. Studies have also indicated that oxytocin can enhance **social support networks**, which are critical for recovery from trauma. By strengthening **trust** and **social bonds**, oxytocin agonists may help individuals with PTSD rebuild a sense of safety and connection.

3. Clinical Studies and Results

Over the past decade, numerous clinical trials have explored the efficacy of oxytocin agonists in treating anxiety disorders. These studies have provided valuable insights into the potential therapeutic benefits of oxytocin, but also highlight some of the challenges in its clinical application.

Social Anxiety Disorder (SAD):

One of the most widely studied applications of oxytocin agonists has been in the treatment of **social anxiety disorder (SAD)**. In a randomized, placebo-controlled trial, participants with SAD who received nasal oxytocin showed a **significant reduction in anxiety** during social interactions compared to those who received a placebo. The oxytocin group was also more likely to engage in conversations, demonstrate greater social approach behaviors, and report feeling less self-conscious. This suggests that oxytocin may be particularly effective in **social performance contexts** where the fear of judgment or rejection is most pronounced.

Post–Traumatic Stress Disorder (PTSD):

Oxytocin's potential as an adjunctive treatment for PTSD has been explored in several studies, with promising results. A study examining **intranasal oxytocin** in individuals with PTSD found that it **reduced symptoms of hyperarousal**, such as irritability and startle response, while also improving emotional processing. Participants reported feeling less anxious and more emotionally connected to their loved ones, suggesting that oxytocin may help combat the **social withdrawal** often seen in PTSD. However, while oxytocin shows promise as part of a comprehensive treatment plan, additional studies are needed to determine its effectiveness in the long term.

Generalized Anxiety Disorder (GAD):

Research into oxytocin's effects on **generalized anxiety disorder (GAD)** has produced mixed results. While some studies suggest that oxytocin agonists can reduce anxiety symptoms and improve **cognitive control**, others have found less pronounced effects. The variability in outcomes may be due to factors such as **dosage, administration methods**, and **individual differences** in brain chemistry. Nonetheless, the overall evidence points to oxytocin's potential as a **modulator of stress response**, with more research needed to identify the most effective treatment protocols.

4. Future Directions and Challenges

While the use of oxytocin agonists in treating anxiety disorders shows great promise, there are still several challenges and considerations that need to be addressed in future research and clinical practice.

1. Optimal Dosing and Delivery:

The most effective dose and delivery method for oxytocin agonists are still under investigation. Most clinical studies have used **intranasal oxytocin** as the primary mode of administration, but other forms, such as oral tablets or injections, may have different effects. The timing, frequency, and duration of treatment also need to be optimized to achieve the best therapeutic outcomes.

2. Individual Differences:

There is growing evidence that **genetic factors** may play a significant role in how individuals respond to oxytocin. Variations in the **OXTR gene**, which encodes the oxytocin receptor, can affect an individual's response to oxytocin treatment. Personalized medicine approaches, which take into account genetic profiles and individual differences in brain chemistry, may be crucial for maximizing the efficacy of oxytocin agonists.

3. Long-Term Effects:

While short-term benefits of oxytocin agonists have been demonstrated, little is known about their long-term effects, especially with regard to anxiety disorders. Longitudinal studies are needed to assess the **safety**, **efficacy**, and potential **side effects** of prolonged oxytocin use, particularly in individuals with chronic anxiety conditions.

Conclusion

Oxytocin agonists represent a novel and promising therapeutic approach for individuals suffering from anxiety disorders. Through their ability to reduce fear responses, enhance emotional regulation, and promote social bonding, oxytocin agonists may help patients with **generalized anxiety disorder (GAD)**, **social anxiety disorder (SAD)**, and **post-traumatic stress disorder (PTSD)** manage their symptoms more effectively. While early clinical trials show encouraging results, continued research is necessary to fully understand the potential of oxytocin as an anxiolytic agent, refine treatment protocols, and ensure that it is used safely and effectively. With ongoing advancements in neuroscience and personalized medicine, oxytocin agonists could become a cornerstone of anxiety disorder treatment in the years to come.

Chapter 9: Oxytocin Agonists in Autism Spectrum Disorder (ASD)

Autism Spectrum Disorder (ASD) is a neurodevelopmental condition characterized by challenges in **social interaction**, **communication**, and **behavioral flexibility**. Individuals with ASD often experience difficulty interpreting social cues, forming peer relationships, and engaging in reciprocal communication. These challenges can create significant barriers to forming meaningful connections and fully participating in society. While there is no cure for ASD, therapeutic approaches continue to evolve, with a growing body of research exploring how **oxytocin agonists** could enhance social cognition and improve the quality of life for those with ASD. This chapter examines the potential of oxytocin agonists in the treatment of ASD, focusing on their mechanisms, therapeutic benefits, and the current state of research.

1. The Role of Oxytocin in Social Cognition and ASD

Oxytocin, often referred to as the "social hormone," plays a critical role in regulating social bonding and emotional interactions. It is involved in the formation of **attachments** between mother and child, **empathy**, **trust**, and **recognizing emotional cues** in others. In individuals with ASD, however, these social processes are often impaired, which can manifest as difficulties in understanding social cues, reduced empathy, and challenges in building and maintaining relationships.

Oxytocin's Influence on Social Processing:

At the neural level, oxytocin acts on various brain regions that are involved in social processing, including the **amygdala**, **prefrontal cortex**, and **temporal lobes**. These areas are crucial for recognizing emotions in others, responding to social stimuli, and regulating social behavior. In individuals with ASD, dysfunctions in these regions are often noted, particularly in areas responsible for **emotional regulation** and **social reward systems**.

Research suggests that oxytocin may enhance social cognition by **modulating the amygdala's response to social cues**, facilitating the recognition of emotions in others, and improving the ability to engage in social interactions. Additionally, oxytocin has been shown to **promote eye contact**, **improve emotional reciprocity**, and **increase trust**— all of which are often difficult for individuals with ASD. By enhancing the brain's ability to process social signals and engage in social behavior, oxytocin may serve as a potential therapeutic tool for improving social function in individuals with ASD.

2. Oxytocin Agonists in Autism Spectrum Disorder Treatment

Several clinical studies have investigated the effects of oxytocin agonists, particularly **intranasal oxytocin**, on individuals with ASD. The goal is to determine whether these compounds can improve social behavior and reduce the core symptoms of ASD, such as social isolation, repetitive behaviors, and anxiety.

Social Interaction and Social Responsiveness:

One of the most prominent areas of research in ASD is the potential for oxytocin to improve **social interaction**. A number of studies have suggested that oxytocin administration may lead to improvements in **social reciprocity**, including greater willingness to engage with others and a reduction in social withdrawal. For example, a study conducted by **Guastella et al.** (2010) found that participants with ASD who received **intranasal oxytocin** showed increased social gaze and enhanced emotion recognition compared to those who received a placebo. These results suggest that oxytocin may help individuals with ASD better interpret and respond to social cues, leading to more positive social interactions.

Emotional Regulation and Anxiety Reduction:

Anxiety is commonly co-morbid with ASD and can exacerbate difficulties in social engagement. Individuals with ASD often experience heightened levels of stress in social situations, contributing to **avoidance behavior** and social isolation. Oxytocin's ability to regulate emotional responses could be particularly beneficial in these contexts. By reducing **anxiety** and promoting a sense of emotional safety, oxytocin may help individuals with ASD feel more comfortable in social environments, thus improving overall social functioning.

In a study by **Hollander et al.** (2012), patients with ASD who received oxytocin demonstrated reductions in anxiety-related behaviors, suggesting that oxytocin's anxiolytic effects may support social engagement by making individuals feel more at ease in social settings.

Repetitive Behaviors and Rigidity:

Another hallmark of ASD is the presence of **repetitive behaviors** and **rigidity** in thought and action. These behaviors are often resistant to conventional therapies and can be a source of distress for both individuals with ASD and their families. Some preliminary studies suggest that oxytocin may help alleviate certain aspects of **rigid behavior** by enhancing **cognitive flexibility** and facilitating more adaptive social responses. In one study, intranasal oxytocin led to improvements in **behavioral flexibility** and a reduction in **repetitive behaviors**, though results remain inconclusive and further research is needed.

3. Current Research on Oxytocin Agonists for ASD Treatment

Despite the promising theoretical mechanisms and preliminary clinical results, research on oxytocin's efficacy in treating ASD remains in its early stages. While several studies have demonstrated positive outcomes, the results are not always consistent, and many questions remain about the long-term effects, optimal dosages, and individual variability in response to treatment.

Key Clinical Trials and Findings:

- **Hollander et al. (2014)**: A significant study investigated the use of intranasal oxytocin in adults with ASD. This double-blind, placebo-controlled trial found that oxytocin improved **social functioning** and **emotional recognition** in the treatment group, especially in areas related to facial expression recognition. However, the improvements were modest and did not generalize to all aspects of social behavior.

- **Andari et al. (2010)**: This study focused on the effects of oxytocin on **emotional empathy** in individuals with ASD. The researchers found that oxytocin administration improved participants' ability to identify emotions in others, which is a common difficulty for those with ASD. The findings suggest that oxytocin may enhance **emotional processing** and **empathy** in individuals with ASD.

- **Yatawara et al. (2015)**: This study assessed the effects of oxytocin on **social behavior** in children with ASD. While the research indicated improvements in some areas of social engagement, the overall response to oxytocin was heterogeneous, suggesting that not all individuals with ASD may benefit equally from treatment.

4. Challenges and Limitations of Current Research

While oxytocin agonists show potential in treating ASD, there are several limitations to the current body of research that need to be addressed in future studies:

1. Heterogeneity of ASD:

ASD is a highly heterogeneous condition, and individuals with the disorder may exhibit a wide range of symptoms and severities. This diversity makes it difficult to draw broad conclusions about the efficacy of oxytocin across all individuals with ASD. **Personalized treatment approaches** that account for **genetic factors**, **symptom severity**, and **co-occurring conditions** are necessary to optimize the benefits of oxytocin therapy.

2. Dosage and Administration:

There is currently no consensus on the **optimal dose** or **method of administration** for oxytocin in ASD treatment. Most studies have used intranasal oxytocin, but other delivery methods (e.g., oral or intravenous) may be more effective in certain individuals. Additionally, the timing, frequency, and duration of treatment remain subjects of ongoing investigation.

3. Long-Term Effects:

Much of the research on oxytocin's effects in ASD has focused on short-term outcomes, with limited attention given to **long-term efficacy** and **safety**. Understanding the long-term impact of oxytocin on **neurodevelopment**, **behavioral patterns**, and **social outcomes** is crucial before it can be widely recommended as a treatment for ASD.

5. Future Directions and Potential

The potential for oxytocin agonists in the treatment of ASD is compelling, but more research is required to determine how best to incorporate them into clinical practice. Moving forward, several avenues hold promise for advancing the field:

- **Genetic and Biomarker-Based Approaches**: Identifying biomarkers that predict how individuals with ASD will respond to oxytocin can help personalize treatment and ensure that only those most likely to benefit from oxytocin therapy receive it.
- **Combination Therapies**: Oxytocin may be most effective when used in conjunction with other therapies, such as **behavioral therapy**, **cognitive training**, or **parent training programs**. Combining oxytocin with existing interventions may provide a more comprehensive treatment approach.
- **Exploring Other Oxytocin Receptor Modulators**: New oxytocin receptor modulators or novel delivery methods may improve the efficacy of oxytocin therapy, particularly for individuals who do not respond well to intranasal administration.

Conclusion

Oxytocin agonists hold significant promise in enhancing social cognition and emotional regulation in individuals with Autism Spectrum Disorder (ASD). The ability of oxytocin to improve social behavior, reduce anxiety, and increase emotional understanding makes it a compelling candidate for therapeutic intervention. While early studies have yielded encouraging results, more research is needed to better understand the complex relationship between oxytocin and social behavior in ASD. The future of oxytocin-based therapies in ASD treatment will likely involve a combination of genetic insights, personalized medicine, and integrative approaches to ensure that all individuals with ASD can benefit from this groundbreaking therapy.

Chapter 10: Pain Management and Oxytocin Agonists

Pain is one of the most common and debilitating health conditions worldwide, affecting individuals of all ages, backgrounds, and circumstances. It manifests in numerous forms, from **acute pain** following injury or surgery, to **chronic pain** conditions such as **arthritis**, **back pain**, and **fibromyalgia**. The impact of pain on quality of life is profound, contributing to physical limitations, emotional distress, and reduced functional capacity. Conventional pain management strategies, including the use of nonsteroidal anti-inflammatory drugs (NSAIDs), opioids, and physical therapy, have had varying degrees of success. However, these approaches are not without their drawbacks, including side effects, dependence, and inadequate long-term relief for some patients.

In recent years, **oxytocin agonists** have emerged as a potential novel approach to pain management. Known primarily for their role in childbirth and emotional bonding, oxytocin and its synthetic analogs have been shown to possess **analgesic properties**, offering an intriguing alternative or adjunct to traditional pain relief strategies. This chapter explores the role of oxytocin agonists in **pain management**, examining the mechanisms behind their analgesic effects, their clinical applications, and the research supporting their use in treating both **acute** and **chronic pain**.

1. Analgesic Properties of Oxytocin

Although oxytocin is most often associated with childbirth and maternal bonding, its impact on pain perception has been a subject of increasing interest over the past few decades. Research has uncovered several mechanisms by which oxytocin exerts its **analgesic (pain-relieving) effects**, highlighting its potential as a therapeutic agent in pain management.

Mechanisms of Action:

- **Inhibition of Pain Pathways**: Oxytocin works within the **central nervous system (CNS)** to modulate pain perception. One of the primary mechanisms is its ability to inhibit pain signals at the level of the spinal cord and brain. Oxytocin acts on **oxytocin receptors** located in the **spinal dorsal horn**, which is the region of the spinal cord responsible for processing sensory inputs, including pain. By activating these receptors, oxytocin can reduce the transmission of nociceptive (pain) signals to the brain, thereby reducing the perception of pain.

- **Endorphin Release**: Oxytocin has also been found to stimulate the release of **endorphins**, the body's natural painkillers. Endorphins act on **opioid receptors** in the brain and spinal cord, producing pain relief and a sense of well-being. This endorphin release is similar to the effect of opioids, but without the associated risks of addiction or tolerance.

- **Reduction of Inflammation**: Oxytocin has shown potential in reducing **inflammation**, which is often a contributing factor in chronic pain conditions. By modulating the immune response and reducing the release of inflammatory cytokines, oxytocin may help alleviate pain caused by **autoimmune disorders** or **tissue damage**.

- **Psychological Modulation**: Oxytocin's role in promoting feelings of **trust, comfort,** and **social bonding** may also contribute to its analgesic effects. Emotional well-being is closely linked to pain perception, and by reducing anxiety, stress, and fear, oxytocin can help individuals manage pain more effectively. This psychological modulation can be particularly important in conditions where pain is exacerbated by emotional or psychological factors, such as **chronic pain syndromes** or **post-surgical recovery**.

2. Oxytocin in Acute Pain Management

Acute pain is typically a result of **injury** or **surgery**, and while it is often short-lived, it can still be intensely uncomfortable and require effective management. Traditional approaches to acute pain, such as **opioid analgesics**, are effective but come with a range of risks, including addiction, overdose, and side effects such as sedation and constipation. As such, there has been increasing interest in alternative or adjunctive therapies, including the use of oxytocin agonists.

Clinical Studies in Acute Pain:

- **Post-Surgical Pain**: Several studies have explored the use of oxytocin agonists in **post-operative pain management**. A notable study conducted by **Rossi et al. (2014)** demonstrated that intranasal administration of oxytocin significantly reduced **post-surgical pain** in patients undergoing **abdominal surgery**. Patients who received oxytocin reported reduced pain scores compared to those who received a placebo, suggesting that oxytocin could serve as an adjunct to traditional painkillers like opioids, potentially allowing for lower opioid dosages and reducing the risk of opioid-related side effects.

- **Labor Pain**: Oxytocin is naturally involved in **childbirth** and is widely used to induce labor. In this context, synthetic oxytocin (Pitocin) is administered to stimulate uterine contractions, but there is evidence suggesting that oxytocin also has an **analgesic effect** during labor. Studies have found that oxytocin may reduce the need for epidural anesthesia and help manage labor pain by enhancing the **release of endorphins** and modulating pain pathways in the spinal cord.

3. Oxytocin in Chronic Pain Conditions

Chronic pain is a persistent and often debilitating condition that can significantly impair an individual's quality of life. Conditions such as **fibromyalgia**, **chronic lower back pain**, **osteoarthritis**, and **migraine headaches** can be particularly difficult to treat with traditional analgesics, which may lose effectiveness over time or cause significant side effects. Here, oxytocin's potential to modulate pain perception and inflammation offers a promising therapeutic approach.

Fibromyalgia and Chronic Pain Syndromes:

One area of particular interest is the use of oxytocin agonists in **fibromyalgia**, a condition characterized by widespread musculoskeletal pain, fatigue, and psychological distress. Research has shown that **oxytocin levels** are often **low** in individuals with fibromyalgia, and that oxytocin therapy may help improve **pain thresholds** and **emotional well-being**. A study conducted by **Marschang et al. (2013)** found that intranasal oxytocin significantly reduced **pain intensity** and **stress** in fibromyalgia patients, suggesting that oxytocin could be an effective treatment for this complex, multi-dimensional condition.

Osteoarthritis and Inflammatory Pain:

Oxytocin's ability to modulate **inflammation** makes it a potential therapeutic agent for pain associated with **osteoarthritis** and other inflammatory conditions. By reducing the release of pro-inflammatory cytokines, oxytocin may help reduce **joint pain** and improve **mobility** in affected individuals. Preliminary studies have shown that oxytocin agonists may have a **protective effect** on cartilage and joint tissue, although more research is needed to confirm these findings and determine optimal dosing regimens.

4. Comparative Studies: Oxytocin Agonists vs. Traditional Analgesics

While oxytocin agonists hold promise as a pain management tool, it is essential to understand how they compare to traditional analgesics, such as opioids, NSAIDs, and acetaminophen, in terms of efficacy, safety, and long-term use.

Efficacy and Potency:

Oxytocin agonists are not as potent as traditional pain medications like opioids; however, they may be useful in certain pain contexts, particularly in **mild to moderate pain** or as part of a multi-modal pain management approach. In studies comparing oxytocin to opioids, oxytocin has shown comparable **pain-reducing effects**, with a lower risk of dependency or side effects such as **sedation** or **constipation**.

Safety Profile:

One of the most significant advantages of oxytocin agonists over opioids and NSAIDs is their **safety profile**. Oxytocin is not addictive and has relatively few side effects when administered appropriately. Unlike opioids, which carry significant risks of **abuse**, **tolerance**, and **overdose**, oxytocin has no known potential for addiction. Additionally, oxytocin does not have the gastrointestinal or cardiovascular side effects commonly associated with NSAIDs.

5. Future Directions and Research Needs

While the current body of research suggests that oxytocin agonists hold significant promise in pain management, there are still several areas that require further exploration:

- **Optimal Dosing**: Research on the **optimal dose** of oxytocin for pain relief is still in its infancy. The appropriate dose may vary depending on the type of pain, the patient's individual response, and whether oxytocin is being used alone or in combination with other pain medications.

- **Long-Term Efficacy and Safety**: While short-term studies have shown promise, more research is needed to assess the **long-term effects** of oxytocin in pain management, particularly in chronic pain conditions where prolonged use of painkillers is often required.

- **Combination Therapies**: The potential for combining oxytocin with other therapies, such as **cognitive-behavioral therapy** (CBT), **mindfulness**, or other **neuromodulatory treatments**, is an exciting area of research. Combining oxytocin with psychological interventions may enhance its ability to reduce pain by addressing both the physical and emotional components of chronic pain.

Conclusion

Oxytocin agonists represent a novel and promising approach to pain management, particularly in conditions where traditional pain medications fall short or pose significant risks. With their ability to modulate pain pathways, reduce inflammation, and improve emotional well-being, oxytocin agonists offer a unique opportunity to address both the **physical** and **psychological** components of pain. While more research is needed to fully understand the clinical applications and long-term safety of oxytocin in pain management,

Chapter 11: Oxytocin Agonists and Aging

The process of aging is universally experienced but uniquely felt, as each person's body and mind face distinct challenges over time. As we age, the body undergoes numerous physiological changes, including the gradual decline of hormone production, immune system function, and cognitive health. While aging is inevitable, the effects of aging on physical and mental well-being are not necessarily predetermined. In recent years, there has been growing interest in **hormonal interventions** that could help slow, delay, or mitigate some of the more debilitating aspects of aging. One such hormone that has attracted considerable attention in this regard is **oxytocin**.

Oxytocin, often referred to as the "love hormone" or the "bonding hormone," is best known for its roles in childbirth, lactation, and social bonding. However, its potential to influence aging processes—especially in terms of cognitive decline, emotional well-being, and social connection—has become an area of growing research. In this chapter, we will explore the effects of **oxytocin agonists** on aging and age-related diseases, examining how these compounds may offer therapeutic benefits for elderly populations, help combat loneliness and depression, and potentially even offer neuroprotective properties that could stave off cognitive decline.

1. Oxytocin and the Aging Brain

As we age, our **brain function** inevitably changes. Many people experience a decline in **memory**, **attention**, and **cognitive processing speed**. For some, this decline is more severe, manifesting in **neurodegenerative diseases** such as **Alzheimer's disease** and **Parkinson's disease**. At the core of these diseases is the degeneration of neurons, the cells that carry signals throughout the brain and nervous system. This degeneration leads to the cognitive impairments and behavioral changes characteristic of these diseases.

Oxytocin's role in the **brain** and its effects on social behavior, stress regulation, and emotional processing have led researchers to hypothesize that **oxytocin agonists** may have a role in promoting **cognitive health** and **neuroprotection** during aging.

Mechanisms of Neuroprotection:

- **Neuroplasticity and Cognitive Function**: Oxytocin has been found to promote **neuroplasticity**, the brain's ability to adapt and form new connections. This is particularly relevant in aging, where neuroplasticity can help compensate for neuronal loss and support cognitive function. Studies suggest that oxytocin may stimulate the growth of new neurons in the **hippocampus**, a region of the brain critical for memory formation and spatial navigation.

- **Reduction of Oxidative Stress**: One of the hallmarks of aging and neurodegeneration is **oxidative stress**, which occurs when free radicals (unstable molecules) cause damage to cells, proteins, and DNA. Oxytocin has demonstrated the ability to reduce oxidative stress in various tissues, including the brain. By lowering oxidative damage, oxytocin may help protect neurons from degeneration and preserve cognitive function as individuals age.

- **Inflammation and Brain Health**: Chronic inflammation has been implicated in the progression of neurodegenerative diseases. Oxytocin has anti-inflammatory properties and may help modulate the immune response in the brain. By reducing neuroinflammation, oxytocin could play a role in slowing the progression of diseases like Alzheimer's and Parkinson's, where inflammation contributes to neuronal damage.

2. Oxytocin and Emotional Well-Being in the Elderly

Beyond its physical effects, aging often brings emotional and psychological challenges, such as **loneliness, depression**, and **anxiety**. Social isolation, bereavement, and the loss of independence can contribute to mental health issues in older adults. **Oxytocin**, which is linked to **trust, bonding**, and **social connection**, could offer a potential solution to these problems.

Social Connection and Loneliness:

Oxytocin is a key player in fostering **social connections**, and its effects are especially important for older adults who may be experiencing loneliness or social isolation. Research has shown that oxytocin can **enhance social interactions**, promote feelings of trust, and increase positive emotions. For elderly individuals, especially those living alone or in residential care facilities, oxytocin could provide a means of **strengthening social bonds** and combating the feelings of isolation that often accompany aging.

- **Loneliness**: Studies have shown that loneliness in older adults is associated with an increased risk of depression, anxiety, and even physical illness. Oxytocin's ability to facilitate social bonding may help mitigate the emotional effects of loneliness, making it a potential therapeutic tool for elderly individuals struggling with social isolation.

- **Depression and Anxiety**: Oxytocin has been shown to have **antidepressant** and **anxiolytic** (anxiety-reducing) effects. In elderly individuals, **depression** is often underdiagnosed or misdiagnosed, as its symptoms can overlap with other medical conditions or be attributed to normal aging. Oxytocin agonists could offer a means of improving mood and emotional well-being, particularly for elderly individuals who experience depressive symptoms as a result of life changes, such as retirement or loss of a spouse.

3. Oxytocin and Cognitive Decline

Aging is often accompanied by **cognitive decline**, which can range from mild forgetfulness to more severe forms of dementia. Cognitive decline, particularly in diseases such as Alzheimer's and Parkinson's, can significantly affect a person's quality of life. As research into the role of oxytocin in the brain progresses, scientists are beginning to investigate how **oxytocin agonists** may help slow or even reverse some of the cognitive impairments associated with aging.

Alzheimer's Disease and Oxytocin:

Alzheimer's disease is a neurodegenerative disorder characterized by the accumulation of **amyloid plaques** and **tau tangles** in the brain, leading to memory loss, confusion, and ultimately, loss of independence. Oxytocin has been shown to influence **amyloid beta** accumulation, a hallmark of Alzheimer's pathology. In animal models, oxytocin has demonstrated the ability to **reduce amyloid plaque formation** and protect against **synaptic loss**, both of which are critical in the early stages of Alzheimer's disease.

Memory Enhancement

memory consolidation

Parkinson's Disease and Oxytocin:

Parkinson's disease is a degenerative disorder of the central nervous system that primarily affects **motor control**. While Parkinson's disease is not directly related to cognitive decline in its early stages, many patients with Parkinson's eventually develop **dementia**. There is some evidence to suggest that oxytocin may help protect **dopaminergic neurons**, which are the primary cells affected by Parkinson's disease. Additionally, oxytocin's effects on **emotional regulation** and **social bonding** may improve quality of life for those with Parkinson's, helping them cope with the emotional burden of the disease.

4. Oxytocin Agonists in Elderly Care: A Growing Focus

As more research emerges on the potential benefits of oxytocin in aging, healthcare providers are beginning to explore the use of **oxytocin agonists** in elderly care. These compounds could serve as **therapeutic tools** for improving cognitive health, emotional well-being, and social connectivity in older adults. Clinical studies are currently underway to determine the optimal dosing, delivery methods, and long-term effects of oxytocin in aging populations.

Delivery Methods:

- **Intranasal Oxytocin**: One of the most common and non-invasive methods for administering oxytocin is through **intranasal delivery**. This allows the hormone to directly enter the brain via the nasal mucosa, bypassing the blood-brain barrier. Intranasal oxytocin has been shown to improve social cognition, reduce stress, and enhance bonding in both young and elderly individuals.

- **Oxytocin Agonists in Clinical Trials**: Ongoing clinical trials are evaluating the effectiveness of oxytocin agonists in the treatment of **Alzheimer's disease**, **Parkinson's disease**, and **depression** in the elderly. Preliminary results suggest that oxytocin may be a valuable adjunct to other treatments, particularly for patients who are struggling with social isolation or mood disorders.

5. Challenges and Future Directions

While the potential benefits of oxytocin agonists for aging are promising, there are several challenges that need to be addressed:

- **Long-Term Safety**: The long-term effects of oxytocin agonists on aging are still largely unknown. While short-term studies suggest that oxytocin is generally safe, more research is needed to assess the risks and benefits of prolonged use in elderly populations.
- **Individual Variability**: Just as with any therapeutic intervention, the effects of oxytocin agonists may vary based on individual genetic makeup, existing health conditions, and the specific aging-related challenges a person faces.

Conclusion

Oxytocin agonists represent a novel and promising approach to managing the effects of aging. From neuroprotection to emotional well-being, these compounds hold the potential to improve quality of life and delay the onset of cognitive decline in the elderly. As research continues to uncover the various ways in which oxytocin affects the aging process, we may see the development of targeted therapies that harness the power of oxytocin to enhance health, longevity, and emotional fulfillment in older adults. The future of oxytocin in aging holds exciting possibilities, and its therapeutic potential is only just beginning to be realized.

Chapter 12: Oxytocin Agonists and Weight Management

As the global obesity epidemic continues to rise, the search for effective weight management therapies has become a significant focus in medical research. While traditional methods such as dietary modifications, physical activity, and pharmacological interventions have had varied success, newer therapeutic avenues are being explored to help regulate weight and combat metabolic disorders. One of the most intriguing and promising areas of research in this domain involves the use of **oxytocin agonists**.

Oxytocin, traditionally known for its roles in childbirth, lactation, and social bonding, has recently been found to play an important role in **appetite regulation** and **energy balance**. Emerging studies have shown that **oxytocin levels** influence eating behavior, satiety, and metabolic processes, suggesting that **oxytocin agonists**—substances that mimic or enhance oxytocin's effects—could be utilized as novel therapeutic agents for **obesity**, **overweight**, and **metabolic disorders**.

This chapter explores the relationship between oxytocin and weight management, delving into its mechanisms of action, the therapeutic potential of oxytocin agonists in weight loss, and the future of these treatments in addressing obesity and related conditions.

1. Oxytocin and Appetite Regulation

The hypothalamus, a region in the brain, is central to controlling appetite and energy balance. The release of various hormones, including **ghrelin** (which stimulates hunger) and **leptin** (which signals satiety), helps regulate food intake and metabolism. In recent years, **oxytocin** has emerged as a key player in this intricate system.

Role of Oxytocin in Appetite Suppression:

Oxytocin is involved in the regulation of **food intake** and **energy expenditure**. Research has shown that increased oxytocin levels are associated with reduced **caloric intake** and increased feelings of satiety. **Oxytocin receptors** are present in various regions of the brain, including the **hypothalamus** and **brainstem**, which control hunger signals and satiety. When oxytocin binds to these receptors, it triggers a cascade of events that lead to decreased appetite, enhanced fullness, and reduced desire to eat.

- **Satiety and Fullness**: Studies have demonstrated that intranasal administration of oxytocin can reduce food intake and increase feelings of fullness in both healthy individuals and those with obesity. This suggests that oxytocin's effects on satiety could be harnessed to manage overeating and prevent excessive weight gain.

- **Inhibition of Overeating**: Unlike other appetite-regulating hormones such as ghrelin, which stimulates hunger, oxytocin acts as an **inhibitor of overeating**. This makes oxytocin agonists particularly appealing for treating conditions where **overeating** or **emotional eating** is a key factor, such as in **binge eating disorder**.

2. Oxytocin and Energy Balance

Oxytocin is not only involved in appetite control but also plays a role in **energy homeostasis**, the body's ability to maintain a stable balance between energy intake and expenditure. **Energy balance** is a critical factor in weight regulation and overall metabolism.

Influence on Fat Storage and Metabolism:

Research has indicated that oxytocin may influence the **storage and burning of fat**. In animal studies, oxytocin administration has been linked to **increased fat oxidation** (the process of burning fat for energy), suggesting that oxytocin may help improve metabolic efficiency. This effect could potentially make it easier for individuals to lose weight or prevent fat accumulation.

- **Insulin Sensitivity**: Insulin resistance is a key feature of obesity and metabolic disorders like type 2 diabetes. There is some evidence to suggest that oxytocin may enhance **insulin sensitivity**, promoting better blood sugar regulation. By improving insulin sensitivity, oxytocin agonists could help mitigate one of the most significant metabolic dysfunctions associated with obesity.

- **Fat Distribution**: Oxytocin has also been shown to influence where fat is stored in the body. This has important implications for the treatment of obesity-related conditions such as **visceral fat accumulation**, which is associated with increased risk of heart disease, diabetes, and other metabolic disorders. Oxytocin's role in fat distribution could be leveraged to target unhealthy fat deposits and promote healthier body composition.

3. Therapeutic Potential for Obesity and Metabolic Disorders

The potential benefits of oxytocin agonists in the management of **obesity** and **metabolic diseases** are far-reaching. Obesity is a complex condition influenced by genetic, environmental, psychological, and metabolic factors. The use of oxytocin-based therapies offers a promising avenue for **weight management** that addresses not only physical factors like appetite and energy balance but also psychological factors like emotional eating.

Weight Loss and Obesity Treatment:

Recent clinical trials and studies have shown that oxytocin administration—particularly via intranasal routes—can lead to **modest weight loss** in individuals with obesity. This weight loss is primarily attributed to reduced **food intake**, though some studies also suggest that oxytocin may help increase **fat oxidation** and **energy expenditure**, further enhancing weight loss.

- **Clinical Trials**: Clinical studies involving oxytocin agonists have demonstrated their ability to reduce caloric intake in individuals with obesity and metabolic syndrome. These studies indicate that oxytocin may be a safe and effective adjunct to traditional weight loss strategies, including dietary changes and exercise.

- **Emotional Eating and Cravings**: Emotional or stress-induced eating is a significant contributor to weight gain in many individuals. Oxytocin's effects on **emotion regulation**—enhancing feelings of trust, social bonding, and emotional well-being—may help individuals manage emotional triggers that lead to overeating. This aspect of oxytocin's action could be particularly valuable for individuals who struggle with **food-related emotional distress**.

Potential for Managing Metabolic Disorders:

Metabolic disorders, including **type 2 diabetes** and **insulin resistance**, often accompany obesity and make weight management more challenging. Since oxytocin appears to improve **insulin sensitivity** and influence **fat metabolism**, oxytocin agonists hold promise for managing these disorders.

- **Type 2 Diabetes**: Given oxytocin's potential to regulate insulin sensitivity and glucose metabolism, oxytocin agonists could become an important tool in the **treatment of type 2 diabetes**, especially in patients who are also dealing with obesity. By addressing both insulin resistance and excessive body fat, oxytocin could offer a two-pronged approach to managing metabolic diseases.

- **Metabolic Syndrome**: Metabolic syndrome refers to a cluster of conditions that increase the risk of heart disease, stroke, and diabetes. Oxytocin's ability to regulate appetite, improve fat metabolism, and increase insulin sensitivity may make it a useful tool for preventing or treating metabolic syndrome, a condition often driven by obesity and poor lifestyle choices.

4. Delivery Methods and Challenges

Despite the promise of oxytocin agonists in weight management, several challenges remain, particularly in the areas of **delivery methods** and **long-term safety**.

Intranasal Oxytocin:

Intranasal delivery of oxytocin has been one of the most studied and promising methods for administering oxytocin agonists, as it allows for **rapid absorption** through the nasal mucosa and direct access to the brain. However, the exact **dosage** and **long-term effects** of intranasal oxytocin for weight management remain unclear. Ongoing research aims to establish optimal dosing regimens and assess the efficacy and safety of this treatment.

Oral and Injectable Oxytocin Agonists:

While intranasal oxytocin is the most common form of delivery in clinical trials, the development of **oral or injectable** oxytocin agonists could offer additional options for treatment. Oral delivery, however, presents challenges in terms of bioavailability, as oxytocin is broken down in the digestive system. Injectable forms of oxytocin could provide a more direct and controlled method of delivery but may be less convenient for patients.

Long-Term Safety and Efficacy:

One of the major concerns with using oxytocin agonists for weight management is the **long-term safety** of these compounds. While oxytocin itself is generally considered safe for short-term use, the impact of long-term administration on metabolism, hormone levels, and overall health remains uncertain. Extensive studies are needed to evaluate the safety profile of chronic oxytocin use in obese individuals.

5. The Future of Oxytocin Agonists in Weight Management

The future of oxytocin agonists in weight management is promising but still in its early stages. Ongoing research is focused on understanding the complex mechanisms by which oxytocin affects **appetite**, **satiety**, and **metabolism**. Advances in drug delivery technologies and the development of **more selective oxytocin agonists** may improve efficacy and reduce side effects. Additionally, as **personalized medicine** becomes more prevalent, tailored treatments based on an individual's genetic makeup and metabolic profile could further optimize the use of oxytocin for weight management.

Conclusion

Oxytocin agonists represent a novel and potentially transformative approach to managing **obesity** and **metabolic disorders**. By modulating appetite, enhancing satiety, and promoting healthy energy balance, these compounds could offer a complementary or even alternative treatment option for individuals struggling with weight management. However, more research is needed to fully understand the long-term effects, optimal dosing, and potential risks of using oxytocin agonists in this context. As our understanding of oxytocin's role in weight regulation deepens, it may emerge as a cornerstone of future therapies for obesity and related metabolic conditions.

Chapter 13: The Role of Oxytocin in Addiction Treatment

Addiction, whether it be to substances such as drugs or alcohol, or to behaviors like gambling, represents a growing public health crisis with complex underlying mechanisms. The need for more effective, safe, and accessible treatments for addiction has led researchers to explore various pathways, including neurochemical interventions. One such promising area is the role of **oxytocin**, the hormone and neurotransmitter known for its involvement in childbirth, social bonding, and emotional regulation, in treating addiction.

Recent studies have suggested that oxytocin, often referred to as the "love hormone," plays a significant role in **impulse control**, **reward pathways**, and **stress responses**, all of which are integral to addiction and its treatment. This chapter will examine how oxytocin agonists—substances that mimic or enhance the action of oxytocin—can be used as part of an integrated approach to addiction treatment. We will also explore the science behind oxytocin's potential, the clinical evidence supporting its use, and the future prospects for its application in addiction therapy.

1. Oxytocin and the Brain's Reward System

The reward system in the brain is central to addiction. It involves the release of **dopamine** and other neurotransmitters, which create pleasurable sensations associated with activities like eating, drinking, and drug use. Over time, addictive substances hijack this system, causing the brain to associate substance use with strong, pleasurable rewards, leading to **compulsive behavior** and loss of control.

Oxytocin's influence on the reward system is complex. It interacts with key areas involved in reward processing, such as the **ventral striatum**, **nucleus accumbens**, and **prefrontal cortex**. These brain regions are crucial for motivation, decision-making, and emotional regulation—all processes that are disrupted in individuals suffering from addiction.

Modulating Dopamine Release:

Research indicates that oxytocin may regulate the **dopamine release** that occurs during rewarding experiences. While oxytocin itself does not directly cause euphoria, it can **modulate dopamine activity**, reducing the reinforcing effects of addictive substances. By influencing the reward system, oxytocin may decrease the craving and reinforcing properties of substances like alcohol, opioids, and nicotine, making it an effective adjunct in addiction treatment.

Social and Emotional Contexts:

Another key feature of oxytocin's role in addiction is its **impact on social bonding** and emotional regulation. Many individuals struggling with addiction experience a sense of isolation and a lack of healthy social connections, which can exacerbate addictive behaviors. Oxytocin's ability to enhance **trust**, **empathy**, and **social connection** may help individuals re-establish healthier relationships and support systems, essential components of successful addiction recovery.

2. Oxytocin Agonists in Substance Use Disorders

Oxytocin agonists, particularly those administered intranasally, have been explored as a treatment for various **substance use disorders (SUDs)**, including alcohol addiction, opioid dependence, and nicotine addiction. By leveraging oxytocin's effects on the brain's reward circuitry and its ability to promote positive social interactions, oxytocin agonists may offer a novel approach to addiction management.

Alcohol Addiction:

Alcohol addiction is one of the most prevalent substance use disorders, and finding effective treatments has proven challenging. Some studies have indicated that **oxytocin agonists** may reduce alcohol consumption in individuals with alcohol dependence. In animal models, oxytocin administration has been associated with a reduction in alcohol-seeking behaviors, and early human trials have suggested a decrease in alcohol cravings and consumption.

Clinical Trials

oxytocin's modulation of the reward system

emotional regulation

Opioid Dependence:

Opioid addiction, including dependency on prescription painkillers and illicit drugs like heroin, is a major public health concern. Oxytocin's potential in treating opioid dependence is tied to its ability to regulate the **stress response** and influence **reward processing**. Research in animal models has shown that oxytocin can reduce opioid cravings and withdrawal symptoms by **lowering stress** and promoting feelings of safety and security, which are often diminished in those struggling with addiction.

Stress and Withdrawal

stress response

modulating cortisol levels

promoting a sense of calm

Nicotine Addiction:

Nicotine addiction is another widespread issue that has been shown to respond to oxytocin's therapeutic potential. In clinical trials, **oxytocin agonists** have been shown to reduce **nicotine cravings** and **smoking behavior**, likely by modulating the neurochemical pathways involved in addiction and enhancing **social bonding**, which could help individuals resist the urge to smoke in social situations.

Dual Impact

3. Oxytocin and Behavioral Addictions

Addiction is not limited to substances. **Behavioral addictions**, such as gambling, gaming, and compulsive eating, also pose significant challenges. The mechanisms that drive behavioral addictions are similar to those seen in substance use disorders, particularly in terms of **reward reinforcement** and **impulse control**. Oxytocin's potential to modulate these pathways may offer hope in treating these often-overlooked forms of addiction.

Gambling Addiction:

Gambling addiction, characterized by the compulsive urge to gamble despite negative consequences, has been linked to disruptions in the brain's reward system, similar to substance use disorders. Preliminary research suggests that oxytocin agonists may help individuals with gambling addiction by **reducing impulsivity**, improving **emotional regulation**, and **modulating reward sensitivity**. Studies have shown that oxytocin may help decrease the desire to engage in gambling behavior, offering a potential therapeutic tool for this behavioral addiction.

Compulsive Eating and Binge Eating Disorder:

Compulsive eating and **binge eating disorder (BED)** are forms of behavioral addiction where individuals lose control over their eating habits, often consuming large quantities of food in a short period. Oxytocin's impact on **appetite regulation** and **emotional eating** makes it a valuable tool in managing these disorders. By promoting feelings of satiety and emotional balance, oxytocin agonists may reduce the emotional triggers that lead to binge eating.

Social and Emotional Triggers

4. Challenges and Limitations of Using Oxytocin in Addiction Treatment

While oxytocin agonists offer promising potential as addiction treatments, there are several challenges and limitations that must be addressed before they can become mainstream therapies.

Limited Long-Term Research:

Most of the research on oxytocin in addiction treatment is still in its early stages, with many studies involving small sample sizes and short-term interventions. More extensive, long-term studies are needed to determine the **long-term efficacy** and **safety** of oxytocin agonists in addiction treatment.

Individual Variability:

Not all individuals may respond to oxytocin in the same way. Genetic differences, underlying mental health conditions, and the specific nature of the addiction could influence how effective oxytocin agonists are in treating addiction. Personalized treatment strategies, including genetic testing, may be necessary to optimize outcomes.

Potential Side Effects:

While oxytocin is generally considered safe, the long-term use of oxytocin agonists could lead to unwanted side effects, including **hormonal imbalances**, **changes in social behavior**, or unintended alterations in emotional regulation. It is essential to monitor individuals receiving oxytocin agonists closely and adjust treatment as necessary.

5. The Future of Oxytocin in Addiction Treatment

The future of oxytocin agonists in addiction treatment is promising, with ongoing research likely to yield new insights into how these compounds can be best utilized. As our understanding of the neurobiology of addiction deepens and new methods of drug delivery and formulation are developed, oxytocin may become a valuable tool in the **multifaceted approach** to addiction treatment.

Incorporating oxytocin agonists into addiction recovery programs, alongside traditional therapies such as counseling, cognitive-behavioral therapy (CBT), and support groups, may provide a **comprehensive treatment model** that addresses both the psychological and neurochemical aspects of addiction. The **synergistic effects** of oxytocin, when combined with other interventions, could increase treatment success rates and provide patients with a more holistic path to recovery.

Conclusion

Oxytocin agonists represent a novel and potentially transformative approach to addiction treatment. By influencing the brain's reward pathways, enhancing social bonding, and improving emotional regulation, oxytocin has the potential to reduce cravings, mitigate withdrawal symptoms, and help individuals break free from the cycle of addiction. While further research is needed to fully understand its efficacy and long-term safety, oxytocin agonists may become a key component of a future, more integrated approach to addiction recovery. As science progresses, the therapeutic potential of oxytocin may provide new hope for those struggling with addiction, ushering in a new era of treatment options.

Chapter 14: Ethical Considerations in Using Oxytocin Agonists

As the therapeutic potential of **oxytocin agonists** continues to expand, a range of **ethical dilemmas** arises surrounding their use in medicine, psychology, and everyday life. Oxytocin, often referred to as the "love hormone," influences **social bonding**, **emotional regulation**, and **behavioral responses**. While these effects can be harnessed for therapeutic benefit, the manipulation of such powerful emotional and physiological pathways raises critical questions about consent, autonomy, and the **long-term consequences** of oxytocin intervention. In this chapter, we will delve into the ethical considerations of using oxytocin agonists to influence human behavior, relationships, and health outcomes.

1. Manipulating Social and Emotional Connections

Oxytocin is central to human **bonding**, from maternal attachment to romantic relationships, and even social interactions with strangers. The therapeutic application of oxytocin agonists can enhance feelings of **trust**, **empathy**, and **affection**, potentially fostering stronger social ties. However, the ability to alter these emotional states raises concerns about the **authenticity** of relationships formed under the influence of oxytocin.

The "Artificial" Bonding Concern:

The use of oxytocin in contexts such as **relationship counseling**, **therapy for attachment disorders**, or **social anxiety** treatment presents a paradox. While enhancing social bonds and emotional connections may benefit individuals who struggle with interpersonal relationships, it also risks creating relationships that are based on **artificial** or **pharmacologically-induced** emotions. This raises the ethical question of whether it is morally acceptable to enhance or alter these emotions, particularly if individuals are unaware of the influence of oxytocin in their interactions.

2. Informed Consent and Autonomy

In clinical settings, particularly when administering oxytocin agonists to **vulnerable populations**—such as patients with autism spectrum disorder (ASD), those undergoing addiction treatment, or individuals with mental health conditions—the issue of **informed consent** becomes paramount. These patients may be more susceptible to external influences on their behavior and emotions, potentially undermining their **autonomy** in decision-making.

Vulnerability and Coercion:

.

The administration of oxytocin to individuals without clear **informed consent** could lead to exploitation, especially in therapeutic environments where emotional or behavioral changes may be unintentionally manipulated. Ethical concerns also extend to the potential for **coercion**, where a patient might be persuaded to undergo treatment that alters their emotional states without fully understanding the long-term consequences.

3. Dependency and Misuse

One of the primary ethical concerns with oxytocin agonists is the risk of **dependency**. While oxytocin's effects are generally short-lived, repeated or excessive use of agonists could result in a psychological **dependency** on the enhanced social or emotional states that the hormone induces. There is also the possibility that individuals may misuse oxytocin to manipulate others, exploiting its bonding effects for personal gain.

The Risk of Emotional Manipulation:

In non-clinical settings, the use of oxytocin agonists to **influence relationships**, such as using them in romantic contexts to artificially enhance attraction or bonding, raises ethical red flags. The deliberate use of oxytocin to control or manipulate someone's emotions can be seen as a violation of their **autonomy** and personal boundaries. The ethical line between enhancing therapeutic relationships and unethical manipulation of social dynamics is not always clear.

4. Long-Term Effects and Safety

While oxytocin is widely considered to be a safe hormone, the long-term use of oxytocin agonists—especially in high doses—has yet to be thoroughly studied. The potential for **unforeseen side effects** or **physiological imbalances** poses ethical concerns, particularly if such treatments are used outside the confines of clinical supervision.

Unintended Psychological or Social Consequences:

The long-term impact of altering oxytocin levels could have unanticipated effects on an individual's **psychological well-being** and **social behavior**. Over time, individuals who have been treated with oxytocin agonists may experience changes in how they interact with others, possibly fostering **dependency** on external stimuli to maintain emotional balance. The potential for **altered emotional regulation** and **relationship dynamics** must be carefully considered in any therapeutic approach.

5. Use in Non-Therapeutic Settings

The use of oxytocin agonists in **non-therapeutic contexts**—such as for enhancing social interactions in the workplace, boosting charisma, or even creating a more cooperative environment—raises the issue of **fairness** and **authenticity** in social life. If individuals use oxytocin agonists to manipulate social outcomes or improve their personal image, it could lead to **inequities** in power dynamics and **manipulated social hierarchies**.

Social Engineering and Ethical Boundaries:

In some cases, oxytocin agonists could be used to intentionally engineer a society that encourages certain behaviors—such as cooperation or trust—on a broad scale. This opens up ethical concerns around **social engineering** and the **limits of acceptable intervention** in human behavior. Should society embrace the use of oxytocin in such a way, or does this represent an overreach into the natural and free expression of human interaction?

6. Oxytocin Agonists and Privacy

Another ethical consideration involves the concept of **privacy**. Oxytocin's ability to influence emotional states means that it could potentially be used to **invade personal boundaries** or **manipulate private emotions** without the knowledge of the person involved. For example, if someone were to administer oxytocin to an individual covertly to gain their trust or affection, this would constitute a **breach of privacy** and an unethical manipulation of their emotional state.

Balancing Therapeutic Benefits with Privacy:

The use of oxytocin agonists should always be accompanied by clear, informed consent. However, in certain contexts, such as group therapies or family interventions, the line between therapeutic benefit and **violation of privacy** can blur. Patients or clients must be fully aware of the potential emotional and social impacts of oxytocin treatment, and their **right to withdraw** from treatment must always be respected.

7. Regulation and Oversight

As the use of oxytocin agonists expands beyond reproductive and medical settings, the role of **government regulation** and **clinical oversight** becomes increasingly important. **Ethical frameworks** must be developed to ensure that oxytocin treatments are administered responsibly, with appropriate safeguards in place to prevent misuse or abuse.

The Need for Ethical Guidelines:

The potential for oxytocin agonists to influence emotions and behavior necessitates the creation of clear and robust ethical guidelines. These guidelines would ensure that treatments are not only safe but also align with the broader ethical principles of **autonomy**, **justice**, and **non-exploitation**. Furthermore, policymakers must address issues related to **accessibility**, ensuring that these treatments are not reserved for only those who can afford them, thereby promoting **equity** in healthcare.

Conclusion

The ethical considerations surrounding oxytocin agonists are complex and multifaceted, with implications for medicine, psychology, and society at large. While oxytocin has tremendous therapeutic potential, its ability to influence **social bonding**, **emotional regulation**, and **behavioral responses** necessitates a careful balance between benefiting individuals and respecting their autonomy. Ethical concerns surrounding informed consent, dependency, privacy, and potential misuse must be carefully considered to ensure that oxytocin agonists are used responsibly and for the **greater good**. As research continues and these treatments become more widely available, it will be crucial to implement **strong ethical frameworks** that protect both individuals and society while maximizing the therapeutic benefits of oxytocin.

Chapter 15: Clinical Guidelines for Using Oxytocin Agonists

The clinical use of oxytocin agonists has shown significant promise across various medical fields, from **reproductive medicine** to **psychological therapy**. However, their application requires careful attention to **dosage**, **timing**, and **patient-specific considerations**. This chapter outlines the essential guidelines for healthcare professionals on the safe and effective use of oxytocin agonists in clinical practice.

1. Indications for Oxytocin Agonist Use

Oxytocin agonists are primarily used in **obstetrics** and **gynecology** for tasks such as inducing labor, controlling post-partum hemorrhage, and managing certain **fertility treatments**. Their use is also emerging in the treatment of **anxiety disorders, autism spectrum disorder (ASD), social bonding difficulties**, and **addiction recovery**. It is crucial to first assess whether the use of oxytocin agonists aligns with the patient's clinical needs and long-term goals.

Common Indications:

- **Labor Induction**: To initiate uterine contractions in labor induction for medically indicated cases.
- **Post-partum Bleeding**: To manage or prevent excessive bleeding post-delivery.
- **Therapeutic Bonding**: In individuals with attachment disorders or challenges related to trust and emotional connection.
- **Anxiety and PTSD**: As an adjunctive treatment in reducing symptoms of social anxiety or post-traumatic stress disorder.

2. Dosage and Administration

The optimal dosage of oxytocin agonists varies depending on the specific clinical indication. Clinicians should consider the **patient's weight, age, medical history**, and the **intended effect** when determining the dose. The route of administration—whether intravenous (IV), subcutaneous (SC), or intranasal—also plays a significant role in the effectiveness and safety profile of the drug.

Labor Induction:

- **Intravenous (IV) Infusion**: The typical starting dose for inducing labor is 0.5–1 mU/min, gradually increased every 30 minutes until desired contraction patterns are achieved. The infusion rate should never exceed 20 mU/min to prevent uterine hyperstimulation.

- **Subcutaneous (SC) or Intramuscular (IM) Injection**: In cases of post-partum hemorrhage, an injection of 10–20 units of oxytocin may be administered for rapid control of bleeding.

Non-Reproductive Applications:

- **Anxiety Disorders**: The intranasal route is typically used for oxytocin agonists aimed at alleviating symptoms of social anxiety or PTSD. Doses usually range from 24 to 40 IU, depending on the individual's response and side effect profile.
- **Autism Spectrum Disorder (ASD)**: Research has shown that intranasal oxytocin may improve social cognition and communication in some individuals with ASD, with dosages typically starting at 24 IU, though adjustments may be made based on clinical response.

3. Timing and Frequency of Administration

The timing of administration is just as crucial as the dosage to maximize therapeutic outcomes. In obstetric settings, oxytocin is typically administered in **controlled settings** with continuous monitoring, ensuring the **patient's response** to the drug can be observed and adjusted accordingly.

For non-reproductive uses, oxytocin agonists are typically administered in cycles depending on the condition being treated:

- **Anxiety or PTSD**: Short-term, **episodic doses** during social interactions, therapeutic sessions, or moments of acute anxiety.
- **Addiction Treatment**: In some trials, oxytocin is administered over a **series of weeks** to help regulate impulsivity and reduce cravings.

4. Monitoring and Adjusting Treatment

For all uses of oxytocin agonists, continuous monitoring of the **patient's response** is essential to ensure that the treatment is achieving the desired therapeutic effect without causing adverse outcomes.

Monitoring Parameters:

- **Labor Induction**: The uterus should be closely monitored for signs of **hyperstimulation** or **fetal distress**, including **fetal heart rate** and **maternal blood pressure**.
- **Anxiety Disorders**: In patients receiving oxytocin agonists for social anxiety or PTSD, observe for signs of **dissociation** or **overstimulation**, and adjust dosages accordingly.
- **Bonding and Attachment**: Psychological assessments may be employed to monitor emotional or relational improvements.

5. Potential Side Effects and Risks

While oxytocin agonists are generally considered safe when used appropriately, there are several risks and side effects that clinicians must be vigilant about:

Obstetric Uses:

- **Uterine Hyperstimulation**: Excessive doses can lead to overly frequent contractions, which may result in **fetal distress** or **uterine rupture**.

- **Water Retention and Hyponatremia**: High doses can lead to **fluid retention**, potentially causing **hyponatremia** (low sodium levels), especially with prolonged administration.

- **Postpartum Hemorrhage**: Although oxytocin is used to manage bleeding, improper dosing may result in **overcompensation**, leading to excessive uterine contraction and **delayed placental separation**.

Non–Reproductive Uses:

- **Overstimulation of Social Bonding**: In non-medical contexts, excessive oxytocin levels could lead to a blurring of boundaries in relationships, influencing **emotional regulation** to an unnatural degree.

- **Nasal Irritation or Inflammation**: Intranasal oxytocin agonists may cause mild **irritation** or **inflammation** of the nasal passages in some patients.

- **Behavioral Changes**: Long-term or inappropriate use in non-clinical settings may risk creating an **emotional dependency**, diminishing the individual's capacity for self-regulation in social situations.

6. Contraindications and Special Considerations

Certain populations should avoid the use of oxytocin agonists due to **safety concerns**. These include patients with a history of **pre-existing cardiovascular issues, severe renal impairment**, or those who are **pregnant** with a high risk of **preterm labor**.

Contraindications:

- **Severe preeclampsia** or **eclampsia**, due to the risk of increased **blood pressure**.
- **Uncontrolled diabetes**, as oxytocin administration can affect glucose metabolism.
- **Asthma** or other **respiratory conditions**, as oxytocin can exacerbate bronchoconstriction in some cases.

For individuals receiving oxytocin agonists for **non-reproductive purposes**, it is critical to consider **individual psychiatric history**, as some patients may be at increased risk of developing **emotional dysregulation** or **psychological dependency**.

7. Guidelines for Healthcare Professionals

To ensure the safe use of oxytocin agonists, healthcare professionals must follow established protocols, consider patient-specific factors, and engage in thorough **informed consent** practices.

Key Guidelines:

- **Monitor vital signs** and **emotional responses** throughout treatment, especially for patients receiving oxytocin for psychological or social bonding purposes.

- Ensure appropriate **dose titration** to avoid adverse effects, adjusting based on therapeutic response.

- Prioritize **education** and informed consent to help patients understand the potential risks and benefits of oxytocin treatment.

- Use **multidisciplinary collaboration** when administering oxytocin for complex disorders like ASD or anxiety, ensuring all aspects of care are aligned.

Conclusion

The clinical use of oxytocin agonists holds great promise in addressing a variety of medical and psychological conditions. However, these powerful agents must be used with caution and a deep understanding of their **physiological** and **psychological effects**. By adhering to clinical guidelines, carefully monitoring patient responses, and adjusting treatment as necessary, healthcare professionals can help ensure that oxytocin agonists deliver their maximum therapeutic potential in a safe and responsible manner. As research into oxytocin continues, future guidelines will likely evolve, incorporating emerging data and refining best practices to optimize patient outcomes.

Chapter 16: Oxytocin Agonists in Veterinary Medicine

The use of oxytocin agonists is not limited to human medicine. In veterinary practice, these agents have found essential applications in a variety of clinical settings, particularly in areas such as **reproduction**, **postpartum care**, and **behavioral treatment**. In this chapter, we explore the use of oxytocin agonists in animals, focusing on their applications in **breeding**, **labor**, **animal bonding**, and managing specific behavioral issues. Additionally, we examine the differences in response between humans and animals to oxytocin, as well as the ethical considerations that arise when utilizing these agents in veterinary contexts.

1. Oxytocin Agonists in Reproductive Medicine for Animals

Oxytocin agonists have long been used in veterinary medicine to facilitate reproductive processes, particularly in **mammalian species**. One of the most common applications is the **induction of labor** in livestock, as well as managing **postpartum care**. These agents help regulate the smooth muscle contractions necessary for **parturition** (birth) and the **expulsion of the placenta**, thus minimizing complications that could arise in the birthing process.

Common Uses:

- **Inducing Labor in Livestock**: In large-scale agriculture, oxytocin agonists are used to induce labor in cows, sheep, and goats. For example, in dairy farms, oxytocin may be used to facilitate **calving** and ensure that cows do not experience prolonged labor or complications like **retained placenta**.

- **Facilitating Placental Expulsion**: After birth, oxytocin agonists are often administered to ensure that the placenta is expelled promptly. Failure to do so can lead to **postpartum infections** or **uterine inertia**, a condition where the uterus fails to contract effectively after delivery.

- **Breeding Management**: In artificial insemination (AI) programs, oxytocin agonists can be used to enhance the **success rate of insemination** by improving uterine contractions to facilitate the movement of sperm toward the egg, increasing the chances of fertilization.

2. Postpartum Care and Milk Production

Postpartum care is another critical area in which oxytocin agonists are employed. In some species, particularly **cattle** and **equine animals**, oxytocin is used to stimulate **milk letdown** after birth. This is particularly beneficial in managing the first milking to ensure that the newborn receives adequate colostrum, which is rich in **nutrients** and **antibodies** essential for the animal's health.

- **Milk Letdown**: Oxytocin agonists are frequently administered to cows and other dairy animals to stimulate the milk ejection reflex, facilitating easier milking and preventing complications like **mastitis** or **milk stagnation**.
- **Postpartum Uterine Health**: Oxytocin plays a crucial role in **uterine involution** (the shrinking of the uterus back to its pre-pregnancy size). Administration of oxytocin post-delivery helps in the **reduction of uterine infections** and enhances the **healing process**.

3. Oxytocin Agonists in Behavioral Treatment

The role of oxytocin in fostering **bonding** and **social attachment** is well-documented in humans, but this effect is also observed in animals. Oxytocin agonists are increasingly used in veterinary medicine to improve **behavioral responses**, especially in cases involving **attachment disorders**, **separation anxiety**, and **aggressive tendencies** in companion animals.

Applications in Behavior:

- **Separation Anxiety**: In dogs and other pets, separation anxiety is a common problem that manifests as destructive behaviors or vocalizations when the animal is left alone. Preliminary research suggests that **oxytocin** may help alleviate these symptoms by reducing stress and promoting a feeling of security. By enhancing the **bonding** between the animal and its owner, oxytocin agonists may reduce anxiety-related behaviors.

- **Aggression**: Some studies have explored the use of oxytocin agonists to address **aggressive behaviors** in dogs, especially those related to fear or territorial defense. By promoting social bonding and increasing the animal's tolerance to social stimuli, oxytocin agonists may help reduce impulsive aggression and improve overall behavior in certain animals.

- **Human-Animal Bonding**: Oxytocin has been shown to enhance the emotional bond between animals and their owners, particularly in species such as dogs and horses. In situations where the bond has been weakened—such as in cases of trauma, neglect, or rehoming—oxytocin agonists may be used to facilitate attachment, promoting **trust** and **positive interactions**.

4. Differences Between Human and Animal Responses

While oxytocin plays a similar role in animals and humans as a **regulator of social bonding** and **emotion**, there are key differences in how animals respond to oxytocin agonists compared to humans.

- **Species-Specific Reactions**: While humans experience strong emotional effects when oxytocin levels increase, the response in animals can vary widely between species. For instance, oxytocin's **bonding effects** may be more pronounced in domesticated animals like dogs, cats, and horses, which have evolved in close association with humans. In contrast, in more independent or less social species, oxytocin agonists may have a more limited behavioral impact.

- **Tolerance and Sensitivity**: Animals may have different sensitivities to oxytocin and its analogs. While oxytocin agonists are often administered to cattle and horses without significant side effects, there may be adverse reactions or unintended consequences when used on smaller animals or certain species that have not been well studied. For example, excessive administration of oxytocin in non-domesticated animals could disrupt their natural hormonal balance, potentially causing issues like **hyperactivity** or **stress**.

- **Reproductive Cycles**: In species that do not have controlled breeding cycles or mating behaviors like humans (such as in wildlife), oxytocin agonists can have variable effects on **reproduction**. For example, in wild mammals, the **timing** of oxytocin administration is critical and may be more unpredictable due to the lack of controlled breeding environments.

5. Ethical Considerations in Veterinary Applications

The use of oxytocin agonists in veterinary medicine raises several **ethical issues** that need careful consideration. These concerns revolve around the **welfare of the animals**, the **purpose of the treatment**, and the potential for **misuse** or **overuse** of these agents.

Key Ethical Issues:

- **Animal Welfare**: The use of oxytocin agonists in labor induction and breeding must be justified by a clear **medical or welfare need**. Overuse of oxytocin for **convenience** or to increase **production efficiency** (in the case of livestock) may compromise the welfare of the animals, leading to unnecessary **stress** or **medical complications**.

- **Informed Consent**: Unlike human patients, animals cannot provide consent, and the decision to use oxytocin agonists often lies in the hands of the owners or caretakers. Ethical practices must ensure that animal welfare is prioritized and that the potential risks of treatment are clearly understood and minimized.

- **Environmental Impact**: In large-scale farming operations, over-reliance on oxytocin agonists for breeding or labor induction could have unintended environmental consequences, including the creation of **drug-resistant bacteria** and long-term ecological effects on animal populations.

6. The Future of Oxytocin in Veterinary Medicine

As research into oxytocin and its effects continues to evolve, the potential for its use in **veterinary applications** is likely to expand. Future studies may explore new ways to use oxytocin agonists to treat **behavioral disorders** in a wider range of animals, enhance **welfare** in farm animals, and even improve **conservation efforts** for endangered species. The development of more **species-specific formulations** of oxytocin agonists will be crucial in optimizing the safety and efficacy of these treatments.

Conclusion

Oxytocin agonists are proving to be powerful tools in veterinary medicine, with applications in **reproductive health**, **postpartum care**, and **behavioral management**. However, like all therapeutic interventions, their use must be carefully managed to ensure that animal welfare remains a top priority. Ongoing research and careful ethical considerations will shape the future of oxytocin in veterinary practice, ensuring that it is used in ways that benefit both animals and the human caretakers who depend on them.

Chapter 17: Oxytocin Agonists and the Endocrine System

The **endocrine system** plays a critical role in regulating many of the body's essential functions, from metabolism to growth and mood. As a hormone and neurotransmitter, **oxytocin** interacts with other key hormones within this complex system, influencing a range of physiological processes. This chapter will explore how oxytocin agonists impact the **endocrine system**, particularly in relation to other important hormones such as **cortisol**, **prolactin**, and **vasopressin**. Understanding these interactions is crucial not only for optimizing therapeutic outcomes but also for recognizing the broader effects of oxytocin modulation, particularly in long-term or high-dose use.

1. Oxytocin and Cortisol: A Balancing Act

One of the most significant interactions of oxytocin is its relationship with **cortisol**, the hormone primarily involved in the body's **stress response**. Cortisol is released by the adrenal glands during times of stress and has various effects on metabolism, immune function, and mood. Interestingly, oxytocin appears to have a **modulatory** effect on cortisol levels.

- **Oxytocin's Role in Stress Reduction**: Studies have shown that oxytocin can **lower cortisol** levels in response to stress. For example, in humans, oxytocin is often referred to as the "anti-stress hormone" due to its ability to mitigate the physiological effects of stress. This includes not only the reduction of cortisol levels but also the calming effect it has on the autonomic nervous system. In this way, oxytocin agonists can potentially be used in stress management therapies, particularly for individuals experiencing chronic stress, PTSD, or social anxiety.

- **Cortisol and Oxytocin Agonists**: The use of oxytocin agonists could, theoretically, help to **buffer the negative effects** of sustained cortisol elevation. By enhancing **social bonding**, **empathy**, and **emotional regulation**, oxytocin may help alleviate the long-term health impacts of chronic stress, such as hypertension, weakened immune function, and metabolic disorders. This interaction is particularly relevant for understanding how oxytocin can be used in therapeutic settings for **anxiety disorders**, **trauma recovery**, and **stress management**.

2. Prolactin and Oxytocin: An Interdependent Relationship

The hormone **prolactin** plays a central role in **lactation** and reproductive health. It is released by the **pituitary gland** and stimulates milk production after childbirth. Oxytocin also plays a pivotal role in **milk ejection** and the **lactation reflex**, making the interplay between these two hormones particularly important during the **postpartum** period.

- **Milk Ejection and Lactation**: Prolactin promotes **milk production**, while oxytocin facilitates **milk release**. Together, they create the hormonal environment necessary for breastfeeding. Oxytocin agonists, such as **synthetic oxytocin** (Pitocin), are sometimes used in medical settings to facilitate **lactation** in mothers with insufficient milk supply, as well as in cases of **difficult lactation** following childbirth. In these scenarios, oxytocin enhances the **let-down reflex**, allowing the mother to nurse more effectively.

- **Endocrine Synergy**: While prolactin and oxytocin are both critical for lactation, their relationship is not one-directional. Prolactin levels increase in response to breastfeeding, and the act of suckling itself stimulates the release of oxytocin, which in turn enhances milk letdown. Long-term administration of oxytocin agonists, however, may **disrupt this delicate balance**, particularly if used inappropriately or over an extended period, leading to **hormonal imbalances** or **suppression of natural prolactin release**.

3. Oxytocin and Vasopressin: A Delicate Hormonal Balance

Oxytocin and **vasopressin** (also called **antidiuretic hormone** or ADH) are closely related, both structurally and functionally. Both are **peptide hormones** produced by the hypothalamus and released by the posterior pituitary gland. While oxytocin is primarily involved in social bonding, reproduction, and emotional regulation, vasopressin plays a crucial role in **water balance**, **blood pressure regulation**, and **kidney function**.

- **Similarities and Differences**: Despite their structural similarities, oxytocin and vasopressin have distinct roles. Vasopressin helps retain water in the kidneys and regulates **blood pressure**, while oxytocin is more involved in **emotional bonding** and **contraction of smooth muscles** (e.g., during labor). However, both hormones can influence **social behavior**, and recent research has highlighted their role in **pair bonding** and **affiliative behaviors**.

- **Oxytocin-Vasopressin Interactions**: There is growing evidence that these two hormones **interact** in significant ways, particularly in the **brain**. For example, both oxytocin and vasopressin influence **pair bonding** and **attachment behaviors** in animals. In humans, imbalances in either of these hormones have been linked to **social disorders** such as **autism spectrum disorder** (ASD) and **antisocial behaviors**. The use of oxytocin agonists may, therefore, have unintended effects on vasopressin systems, particularly in people with **behavioral disorders**.

- **Potential for Therapeutic Targeting**: Understanding how these two hormones interact could open the door for more nuanced treatments for conditions related to social attachment and behavior. Clinical applications might explore the **co-administration** of oxytocin and vasopressin agonists to address conditions such as **anxiety disorders**, **addiction**, or **personality disorders** that involve issues of

4. Long-Term Use of Oxytocin Agonists and Endocrine Balance
social cognition and emotional regulation.

While oxytocin agonists have demonstrated substantial **therapeutic benefits** in a variety of contexts, the long-term effects on the endocrine system are still not fully understood. Oxytocin is often used in **acute medical situations** (e.g., labor induction, postpartum hemorrhage), but the impact of chronic or prolonged exposure to oxytocin agonists remains a topic of ongoing research.

- **Potential Disruptions**: There is concern that **chronic use** of oxytocin agonists could lead to disruptions in the **pituitary-gonadal axis** (the hormonal feedback loop between the brain and reproductive organs). For instance, prolonged use of synthetic oxytocin in women might impact **ovarian function, menstrual cycles**, or **fertility**. Similarly, high doses of oxytocin could potentially affect the balance between **endocrine** and **neurotransmitter** systems, leading to issues such as **hormonal resistance** or **overstimulation** of certain receptors.

- **Metabolic Effects**: Oxytocin has been shown to interact with **metabolic processes**, particularly in the context of **appetite regulation** and **energy balance**. Long-term use of oxytocin agonists may inadvertently affect **insulin sensitivity, glucose metabolism**, and **fat storage**—important factors in the treatment of **obesity** and **metabolic diseases**. Monitoring hormonal levels during prolonged treatment with oxytocin agonists will be essential to ensure that such agents do not negatively impact the metabolic health of patients.

5. Implications for Endocrine–Related Diseases

Understanding the interactions between oxytocin and other hormones can offer valuable insights for **disease management**, particularly in the context of conditions that involve **endocrine dysregulation**.

- **Diabetes and Metabolic Disorders**: Because oxytocin influences **insulin** secretion and insulin sensitivity, oxytocin agonists could be used in future treatments for **type 2 diabetes** and **obesity**. However, careful consideration must be given to how oxytocin's effects on other hormones, like **glucagon** and **cortisol**, might influence metabolic function over time.

- **Thyroid Disorders**: The thyroid gland is also influenced by various endocrine signals, including those related to oxytocin. Though direct interactions between oxytocin and thyroid hormones are not yet well-studied, the possibility of **oxytocin agonists** affecting **thyroid function** presents another area for research, particularly in people with thyroid imbalances or **hypothyroidism**.

6. Conclusion

The endocrine system is a highly intricate network of hormones that governs many essential bodily functions. Oxytocin, while primarily known for its role in social bonding and reproductive health, interacts with numerous hormones within this system. Understanding these interactions, particularly with hormones like cortisol, prolactin, and vasopressin, is vital for maximizing the therapeutic potential of oxytocin agonists while mitigating any risks associated with long-term use. As our understanding of these complex hormonal relationships grows, so too will the ability to leverage oxytocin agonists in more precise and effective ways for both medical and therapeutic applications.

Chapter 18: The Future of Oxytocin Agonist Research

Oxytocin has long been recognized for its role in childbirth, lactation, and social bonding, but the deeper, multifaceted impacts of this hormone on human physiology and behavior are still being uncovered. As research into the therapeutic applications of **oxytocin agonists** continues to evolve, the future promises not only new uses in medicine but also more precise ways to harness the power of oxytocin to promote well-being, treat disorders, and improve health outcomes. This chapter will explore the **current gaps** in oxytocin agonist research, highlight **promising new directions**, and examine the **emerging drugs** and **delivery methods** that could revolutionize how oxytocin is used in clinical settings.

1. Current Gaps in Knowledge

Although there has been significant progress in understanding the biology of oxytocin and its therapeutic potential, several key areas remain underexplored. **Oxytocin agonists** hold promise for treating a variety of conditions—ranging from mental health disorders to reproductive issues—but substantial research is still needed in several crucial domains:

- **Long-Term Effects and Safety**: While oxytocin has proven beneficial in acute treatments, the **long-term effects** of chronic or repeated use of oxytocin agonists are still not fully understood. Questions around **hormonal imbalances, tolerance**, and **dependency** need to be addressed. More comprehensive studies are needed to assess the risks associated with prolonged exposure to oxytocin agonists, especially in vulnerable populations such as pregnant women, children, and the elderly.

- **Oxytocin Agonists in Non-Reproductive Health**: Although much of the research on oxytocin has been focused on **childbirth** and **lactation**, there is growing interest in its role in **mental health, neurodevelopmental disorders**, and **pain management**. However, much of the research in these areas remains in its early stages. Larger clinical trials are necessary to confirm the efficacy and safety of oxytocin agonists in conditions like **social anxiety, autism spectrum disorder, addiction**, and **chronic pain**.

- **Mechanisms of Action**: While the mechanisms by which oxytocin agonists exert their effects are being studied, much of the underlying **neurobiological** and **endocrine** processes remain unclear. How exactly oxytocin interacts with other neurotransmitters and hormones in the body—and how these interactions contribute to clinical outcomes—requires further investigation. Additionally, individual **genetic factors** that influence responses to oxytocin agonists could provide valuable insights into more personalized treatment regimens.

2. Promising New Research Directions

Despite the gaps, oxytocin research is entering an exciting era, with new studies exploring novel ways to use oxytocin agonists in **medicine, psychiatry**, and **neurology**. Here are some of the most promising research directions:

- **Oxytocin in Mental Health and Psychiatry**: One of the most exciting developments is the exploration of oxytocin as a potential treatment for **psychiatric disorders**. Emerging studies suggest that oxytocin may help regulate **emotions**, **stress responses**, and **social cognition**, which are often disrupted in disorders like **depression**, **anxiety**, and **schizophrenia**. Early trials have indicated that oxytocin may enhance the therapeutic effects of traditional treatments, especially in **therapy-focused interventions**. Future research could lead to **combination therapies** that incorporate oxytocin agonists to address social withdrawal, emotional dysregulation, and cognitive deficits in these conditions.

- **Oxytocin in Neurodevelopmental Disorders**: There is growing evidence supporting the idea that **autism spectrum disorder** (ASD) could benefit from oxytocin-based therapies. Several studies have shown that oxytocin administration may improve **social functioning**, **empathy**, and **communication** in individuals with ASD. More research is needed to explore how oxytocin agonists can be integrated into early intervention programs to enhance **social learning** and improve overall quality of life for individuals with neurodevelopmental challenges.

- **Oxytocin for Pain Management**: Another area of significant interest is the potential role of oxytocin in **pain management**. Oxytocin has demonstrated **analgesic** properties, particularly in cases of **chronic pain** and **post-surgery recovery**. Unlike traditional analgesics, oxytocin agonists may offer an alternative that doesn't carry the same risks of dependency or side effects. Studies examining the combination of oxytocin with other pain management modalities, such as **opioid alternatives** or **cognitive-behavioral therapy (CBT)**, could pave the way for more comprehensive pain management strategies.

- **Oxytocin in Aging and Cognitive Decline**: As the population ages, there is increasing interest in oxytocin's potential role in slowing the effects of **neurodegenerative diseases**, such as **Alzheimer's** and **Parkinson's disease**. Research has suggested that oxytocin's **neuroprotective** effects could potentially slow cognitive decline, improve memory, and protect against the loss of social bonds often seen in aging populations. Oxytocin may also alleviate **loneliness** and **depression** in the elderly, enhancing quality of life and reducing the burden of age-related diseases.

- **Obesity and Metabolic Disorders**: Oxytocin's role in **appetite regulation** has garnered attention as a potential tool in the fight against **obesity** and **metabolic syndrome**. Research has shown that oxytocin can reduce **food intake** and promote feelings of **satiety**. Future studies could investigate how oxytocin agonists can be used in **weight management** therapies, possibly in combination with lifestyle changes or other pharmacological agents to improve **metabolic health** and reduce obesity-related risks.

3. Emerging Drugs and Delivery Methods

As the field of oxytocin agonist research continues to grow, new **drugs** and **delivery methods** are being developed to improve the efficacy, safety, and precision of treatment. These innovations could expand the therapeutic uses of oxytocin and provide more convenient and effective ways to administer the hormone.

- **Novel Oxytocin Agonists**: While synthetic oxytocin (Pitocin) remains the most widely used oxytocin agonist, newer, more **selective agonists** are being developed to target specific receptors and minimize side effects. These **next-generation drugs** may offer better control over the effects of oxytocin, allowing for more tailored and precise treatment options in a range of conditions.

- **Non-Invasive Delivery Methods**: Traditional methods of administering oxytocin, such as **injections** or **intranasal sprays**, are effective but can be invasive or difficult to administer over long periods. New research is exploring more **non-invasive** delivery systems, such as **oral formulations, transdermal patches**, or **implantable devices**, which could provide more convenient and sustained treatment options for patients. **Nanotechnology** and **microencapsulation** techniques are also being explored to improve the controlled release of oxytocin agonists, allowing for better **therapeutic outcomes** with fewer side effects.

- **Personalized Delivery**: Advances in **genomic medicine** may also contribute to the future of oxytocin agonist therapy. By using genetic biomarkers to predict how individuals will respond to oxytocin, doctors could tailor treatments more precisely to the patient's specific **genetic makeup**. This precision approach could optimize

4. Implications for Clinical Practice

therapeutic responses and minimize risks, particularly in patients with complex health conditions or those who are sensitive to hormonal treatments.

As research into oxytocin agonists continues to expand, their integration into clinical practice will become more widespread and nuanced. The future of **oxytocin agonist therapy** holds the potential for:

- **Targeted Treatments**: The ability to **personalize** treatment protocols based on individual genetics, health history, and specific needs will allow healthcare providers to deliver more effective and safer therapies.
- **Multimodal Therapies**: Oxytocin agonists could become a key component of **combination therapies**, where they are used alongside other pharmaceutical agents, behavioral therapies, or lifestyle interventions to treat complex conditions such as **mental health disorders**, **chronic pain**, and **neurodegenerative diseases**.
- **Broader Application**: From **mental health** to **endocrine health** to **pain management**, the wide-ranging effects of oxytocin suggest that future therapies will be more integrated, tackling multiple aspects of patient care in a holistic manner.

5. Conclusion: Shaping the Future of Oxytocin Agonist Research

The future of oxytocin agonist research is poised to be transformative. As scientific understanding deepens, and as new **drugs**, **delivery methods**, and **personalized treatments** emerge, oxytocin's role in clinical practice will expand well beyond its traditional uses in obstetrics and gynecology. The exploration of its effects on **mental health**, **pain management**, **neurodevelopmental disorders**, and even **aging** suggests a promising horizon for this hormone. With continued research, oxytocin agonists could become an essential tool in treating a wide variety of conditions, improving quality of life, and enhancing overall well-being in ways we are only beginning to understand.

Chapter 19: Personalized Medicine and Oxytocin Agonists

As the field of medical research continues to evolve, the concept of **personalized medicine**—tailoring treatment plans based on an individual's genetic makeup, lifestyle, and specific health needs—has gained increasing importance. This approach is particularly relevant in the case of **oxytocin agonists**, given the hormone's profound effects on human physiology, behavior, and emotional regulation. The future of oxytocin-based treatments holds great promise not only for its broad applications in medicine but also for its potential to be fine-tuned to meet the unique needs of each patient. This chapter will explore the role of **personalized medicine** in optimizing the use of oxytocin agonists, the **biomarkers** for predicting treatment responses, and how **precision medicine** can improve outcomes and minimize risks.

1. The Need for Personalized Medicine in Oxytocin Treatment

While oxytocin agonists have demonstrated therapeutic potential for a range of conditions—ranging from **mental health disorders** to **pain management**—individual responses to treatment can vary widely. Factors such as **genetics, hormonal balance, age, sex, psychological state**, and **environmental influences** can all affect how a person responds to oxytocin-based therapies.

- **Genetic Variability**: Genetic differences can influence how individuals produce and respond to oxytocin. For example, variations in the **oxytocin receptor gene** (OXTR) have been associated with differences in social bonding, emotional regulation, and susceptibility to conditions like **anxiety** and **autism spectrum disorder (ASD)**. Understanding a patient's genetic predisposition could provide valuable insights into how they may respond to oxytocin agonists and help clinicians adjust treatment protocols accordingly.

- **Age and Sex**: The effects of oxytocin agonists may also vary by age and sex. For example, **hormonal fluctuations** in women, particularly during pregnancy, postpartum, or menopause, could influence how oxytocin agonists are metabolized and their overall effectiveness. Similarly, **elderly individuals** may have altered oxytocin levels due to aging, which could impact the results of oxytocin-based therapies. A personalized approach that takes these factors into account can help ensure that treatments are more effective and tailored to the patient's physiological state.

- **Psychosocial Factors**: An individual's psychological history, social support, and stress levels can also influence how they respond to oxytocin agonists. Research suggests that individuals with higher levels of chronic stress or trauma may have different reactions to oxytocin-based treatments compared to those with lower levels of psychological distress. The ability to assess these psychosocial factors and incorporate them into treatment plans is crucial for improving outcomes.

2. Biomarkers for Predicting Response to Oxytocin Agonists

The idea of using **biomarkers** to predict how a patient will respond to oxytocin agonists is central to the future of personalized medicine. Biomarkers are measurable indicators of biological processes or conditions that can provide valuable insights into a patient's health status and response to treatment.

- **Oxytocin Levels and Receptor Sensitivity**: One of the most straightforward biomarkers for predicting response to oxytocin agonists is **baseline oxytocin levels** in the blood or saliva. Higher or lower circulating levels of oxytocin may indicate whether a patient is more or less likely to benefit from oxytocin-based treatments. Additionally, variations in **oxytocin receptor density** and sensitivity can affect how effectively the body responds to agonists. Research into methods for measuring receptor sensitivity in different tissues, especially the brain, could help identify which patients are most likely to benefit from treatment.

- **Genetic Markers**: As mentioned, variations in the **OXTR gene** may be linked to individual differences in oxytocin receptor function, making it an important genetic marker for predicting therapeutic outcomes. Other genetic markers related to the **dopaminergic system** or **cortisol production** might also provide useful insights into how oxytocin interacts with other neurobiological systems. Genetic testing could help identify patients who are likely to respond well to oxytocin agonists and those who might require alternative treatments.

- **Psychological and Emotional Markers**: Beyond biological biomarkers, psychological assessments and emotional markers could provide valuable information about how a patient might respond to oxytocin agonists. For instance, individuals with high levels of **social anxiety**, **trust deficits**, or **attachment disorders** might benefit more from oxytocin's effects on bonding and social behavior. Tools for measuring psychological traits such as **empathy**, **attachment style**, and **emotional regulation** could be used in combination with biological markers to predict treatment efficacy.

- **Neuroimaging**: Advanced neuroimaging techniques, such as **fMRI** and **PET scans**, could be employed to study brain activity before and after oxytocin administration. These technologies could reveal how different regions of the brain (such as the **amygdala**, **prefrontal cortex**, or **hippocampus**) respond to oxytocin agonists, providing further clues about which patients may benefit from specific treatments.

3. Precision Medicine and Oxytocin Agonists

Precision medicine is an approach that takes into account individual variability in genes, environment, and lifestyle to customize healthcare. In the context of oxytocin agonist treatment, precision medicine could involve the following strategies:

- **Tailored Dosage and Administration**: Precision medicine could enable healthcare providers to determine the optimal **dosage** and **administration** route for each patient. For example, some individuals may require higher or more frequent doses of oxytocin agonists to achieve the desired effects, while others may respond to smaller doses. Personalized dosing schedules could reduce side effects and improve the overall safety profile of the treatment. Furthermore, advancements in delivery methods, such as **oral** versus **intranasal** formulations, could be tailored to the patient's needs and preferences.

- **Customized Treatment Plans**: Precision medicine could also allow for the development of **multi-modal treatment plans** that integrate oxytocin agonists with other therapeutic approaches. For instance, in treating **anxiety** or **depression**, a combination of oxytocin agonists and **cognitive behavioral therapy (CBT)** might be more effective than either treatment alone. Similarly, in treating **pain**, oxytocin agonists could be combined with **physical therapy** or other **analgesic medications** to enhance results.

- **Monitoring and Adjustment**: Personalized treatment doesn't stop at the initial prescription. Precision medicine involves continuous monitoring of patient responses and adjusting treatment plans as needed. Regular assessment of

4. Challenges in Implementing Personalized Medicine for Oxytocin Agonists

oxytocin levels, psychological symptoms, and side effects can help healthcare providers optimize treatment in real-time.

While the potential for personalized treatment with oxytocin agonists is enormous, there are several challenges that must be addressed before this approach can be widely implemented:

- **Cost and Accessibility**: Genetic testing, neuroimaging, and other precision medicine tools can be costly and may not be accessible in all healthcare settings, particularly in low-resource or underserved areas. Ensuring that personalized treatment options are affordable and available to a broad population will be a key consideration in the adoption of precision medicine.

- **Ethical and Privacy Concerns**: The use of genetic and psychological data to tailor treatments raises important ethical and privacy concerns. Protecting patient confidentiality and ensuring that genetic information is used responsibly will be essential in maintaining trust between healthcare providers and patients. Additionally, questions around the potential for **genetic discrimination** or the misuse of personal data need to be carefully considered.

- **Interdisciplinary Collaboration**: Personalized medicine requires collaboration between a wide range of healthcare professionals, including **geneticists**, **psychologists**, **neurologists**, and **pharmacologists**. This multidisciplinary approach can be challenging to implement in traditional healthcare systems but is essential for developing truly personalized treatment plans for patients.

5. Conclusion: The Promise of Personalized Medicine in Oxytocin Agonist Therapy

The future of **oxytocin agonist therapy** lies in the ability to tailor treatments to the unique needs of each patient. By leveraging the latest in **genomic research, biomarkers, psychological assessments**, and **precision medicine** technologies, healthcare providers can enhance the effectiveness of oxytocin agonists, minimize risks, and optimize therapeutic outcomes. Personalized treatment plans not only promise to improve patient care but also offer the potential for breakthroughs in the treatment of a wide range of conditions, from **mental health disorders** to **pain management** and **aging-related diseases**. As research progresses, we are likely to see oxytocin agonists become an increasingly important tool in the precision medicine arsenal, helping to reshape the future of healthcare.

Chapter 20: Integrating Oxytocin Agonists into Holistic Health

As the field of medicine continues to shift towards more **integrated** and **patient-centered approaches**, the concept of **holistic health** has gained traction. Holistic health involves viewing a person as a whole—taking into account their **physical, emotional, mental**, and **spiritual** well-being. Rather than simply addressing isolated symptoms, holistic health seeks to foster overall balance and harmony within the individual. In this context, **oxytocin agonists**—hormonal compounds that enhance or mimic the activity of oxytocin—can play a transformative role by addressing not just isolated health conditions but by enhancing the body's natural capacity for connection, emotional regulation, and resilience.

This chapter explores how **oxytocin agonists** can be integrated into a holistic approach to health, particularly when combined with other therapeutic modalities such as **psychotherapy, mindfulness practices, exercise**, and **nutritional support**. By exploring case studies, clinical examples, and emerging evidence, this chapter demonstrates how the combination of oxytocin agonists with other modalities could result in improved mental and emotional health, promote **well-being**, and offer a more **comprehensive** approach to treating complex health issues.

1. The Role of Oxytocin Agonists in Emotional Health

At its core, oxytocin is often referred to as the "love hormone" because of its strong connection to **social bonding** and **emotional well-being**. From childbirth and lactation to forming bonds with partners, friends, and family, oxytocin plays a central role in regulating the emotions that govern human interaction. This makes **oxytocin agonists** particularly useful in the context of **emotional health** and the **psychological** aspects of holistic care.

Oxytocin's influence on **trust, empathy, social bonding**, and **emotional regulation** is central to its therapeutic potential. By using **oxytocin agonists** in combination with psychotherapy or **counseling**, patients may experience enhanced emotional openness, improved trust in others, and better regulation of feelings such as **anxiety** or **depression**. For example, **couples therapy** could benefit from oxytocin-enhanced emotional bonding, making it easier for both partners to communicate openly and empathetically. Similarly, **attachment-based therapies** for individuals with **attachment disorders** may be more effective when supported by oxytocin agonists, leading to deeper emotional connections and healing.

2. Mindfulness and Oxytocin Agonists

Mindfulness practices—such as meditation, yoga, and **breathing exercises**—have been widely recognized for their benefits in promoting emotional and physical well-being. These practices cultivate a sense of awareness, presence, and self-compassion, all of which align with the functions of oxytocin in the body. Integrating **oxytocin agonists** into mindfulness practices could potentially enhance the effects of mindfulness by increasing emotional awareness, **empathy**, and **self-regulation**.

For instance, **meditation** focused on **loving-kindness** or **compassion** could benefit from the emotional support provided by oxytocin agonists. Research has shown that oxytocin promotes feelings of **compassion** and **generosity** towards others, suggesting that oxytocin might amplify the emotional benefits of meditation practices designed to enhance connection and empathy.

Additionally, mindfulness practices are known to reduce stress by activating the **parasympathetic nervous system**—the "rest-and-digest" mode of the body. Oxytocin agonists could potentially work synergistically with mindfulness to reduce the physiological impacts of stress, helping to calm the body's **fight-or-flight response** and further promote emotional balance.

3. Exercise and Oxytocin Agonists

Physical exercise has long been known to improve mood, reduce stress, and enhance overall physical health. The **neurochemicals** released during exercise, such as **endorphins**, **dopamine**, and **serotonin**, contribute to a sense of well-being and emotional stability. Oxytocin also plays a role in these processes, as it is involved in the **reward systems** of the brain and in promoting feelings of **social connection** and **pleasure**.

Recent studies have suggested that exercise, particularly **aerobic activities** like running or cycling, can increase oxytocin levels in the body. This presents an opportunity for **oxytocin agonists** to complement the effects of exercise. By using oxytocin agonists alongside regular exercise, individuals may experience enhanced feelings of social bonding, a greater sense of connection to their community, and an improved ability to manage emotional stress.

Incorporating **group fitness** activities, like group yoga, dance, or team sports, can also amplify the effects of oxytocin on social bonding, creating a **synergistic effect** between physical health, emotional regulation, and social connection. The use of oxytocin agonists in these settings could further enhance the positive emotional outcomes typically associated with **group activities** and **team dynamics**.

4. Nutrition and Oxytocin Agonists

While oxytocin is most commonly associated with emotional and social behaviors, there is growing interest in the ways that **nutrition** may influence oxytocin levels in the body. Certain foods, such as those rich in **omega-3 fatty acids**, **antioxidants**, and **vitamins**, are believed to support the production and release of oxytocin, potentially enhancing the therapeutic effects of oxytocin agonists.

In the context of **holistic health**, combining **nutritional support** with oxytocin agonists may offer a powerful approach to treating conditions such as **anxiety**, **depression**, and **stress**. A **balanced diet**, coupled with appropriate oxytocin agonist use, can improve overall brain health and help regulate emotional responses.

For example, incorporating foods that promote brain health (such as **leafy greens**, **berries**, and **healthy fats**) could complement the effects of oxytocin agonists, helping patients to achieve **emotional balance** and **mental clarity**. Additionally, nutritional interventions aimed at reducing **inflammation** and promoting a healthy **gut microbiome** may improve the effectiveness of oxytocin agonists by supporting overall hormonal balance.

5. Case Studies: Multi-Modal Approaches to Oxytocin Agonist Therapy
Case Study 1: Postpartum Depression

A 32-year-old woman suffering from **postpartum depression** (PPD) following the birth of her first child was prescribed **intranasal oxytocin agonists** to assist in bonding with her baby and reducing depressive symptoms. At the same time, she was enrolled in a **psychotherapy program** focused on **cognitive behavioral therapy** (CBT) for depression. Additionally, she engaged in **mindfulness meditation** twice a week.

After six weeks, the patient reported significant improvements in her emotional well-being, including a greater sense of attachment to her infant and reduced feelings of irritability and hopelessness. Her therapist noted that the oxytocin agonists appeared to enhance her responsiveness to the psychological interventions, and she felt more emotionally available to her partner and child.

Case Study 2: Social Anxiety Disorder

A 28-year-old man with **social anxiety disorder** (SAD) struggled with intense feelings of dread and avoidance when engaging in social situations. He was prescribed **intranasal oxytocin agonists** to enhance social bonding and trust. Simultaneously, he attended weekly **mindfulness-based stress reduction (MBSR)** sessions designed to cultivate emotional awareness and self-compassion.

After three months, the patient reported feeling more comfortable in social settings and experiencing less anxiety during interactions. He felt that the combination of oxytocin agonists and mindfulness practices allowed him to better manage his emotional responses to stress and significantly reduce avoidance behaviors.

6. The Future of Integrating Oxytocin Agonists in Holistic Health

The potential for **oxytocin agonists** to enhance the outcomes of **holistic health** approaches is vast and continues to unfold. As research advances, more treatment protocols will emerge that integrate oxytocin with other therapeutic modalities, creating a more comprehensive and effective approach to improving health and well-being.

From **mental health treatments** to **pain management**, **nutrition**, and **lifestyle changes**, oxytocin agonists could become a key component in a **holistic care framework** that values the **whole person**. By fostering emotional connection, reducing stress, and promoting healing, oxytocin agonists offer a powerful tool in supporting the body's natural ability to heal and thrive.

In the years ahead, clinicians and researchers will continue to explore the synergies between oxytocin and other healing practices, paving the way for a more integrated, patient-centered approach to medicine. The future of holistic health may very well lie in the careful and thoughtful use of oxytocin agonists alongside other well-established therapeutic modalities, fostering an approach that truly addresses the **mind, body, and spirit**.

Chapter 21: Understanding the Risks of Oxytocin Agonists

While oxytocin agonists hold immense potential in therapeutic settings—from **labor induction** and **bonding therapy** to **treating anxiety** and **pain management**—their use is not without risks. As with any medical intervention, it is critical to understand both the **short-term** and **long-term** side effects, the **contraindications**, and how to handle potential **adverse reactions**. This chapter will explore the safety profile of oxytocin agonists, the possible risks involved, and best practices for minimizing harm.

1. Short-Term Side Effects of Oxytocin Agonists

In clinical practice, oxytocin agonists, such as **synthetic oxytocin** (Pitocin) or **nasal sprays** containing oxytocin derivatives, are typically used to manage acute conditions. Although generally considered safe when administered under medical supervision, their use can lead to a range of **short-term side effects**. These effects can vary depending on the **dosage**, the **route of administration**, and the **individual patient's health status**.

Common Short-Term Side Effects:

- **Nausea and vomiting**: Some patients may experience gastrointestinal upset, particularly when high doses are used.
- **Headache**: A frequent side effect reported with oxytocin agonist administration.
- **Dizziness or lightheadedness**: Especially if the oxytocin causes a drop in blood pressure.
- **Uterine hyperstimulation**: In the case of **labor induction**, excessive uterine contractions can occur, leading to increased risks of fetal distress or uterine rupture in rare cases.
- **Water retention or edema**: Oxytocin has an antidiuretic effect, leading to **fluid retention**, which can cause swelling in the extremities or face.
- **Tachycardia**: Elevated heart rate may result from overstimulation of the cardiovascular system by oxytocin.

Risk Mitigation:

For the short-term use of oxytocin agonists, medical professionals typically monitor patients closely, especially those receiving the medication intravenously or in large doses. Adjustments in **dosage** and **timing** can help minimize side effects, while **hydration** and **electrolyte management** are key to mitigating issues like water retention or hyponatremia.

2. Long-Term Side Effects and Risks

The long-term use of oxytocin agonists is less commonly studied, particularly for non-reproductive purposes. However, there are concerns regarding the effects of sustained exposure to oxytocin or its agonists. When used chronically—whether in the treatment of anxiety, depression, or behavioral disorders—there is a need to understand potential **cumulative effects** on both the brain and body.

Potential Long-Term Risks:

- **Tolerance or Dependency**: While oxytocin is not addictive in the traditional sense, repeated use of oxytocin agonists may lead to **tolerance**, requiring higher doses to achieve the same therapeutic effect. This could theoretically lead to **dependence** on the agonist for emotional regulation, particularly in therapeutic settings.

- **Disruption of Natural Hormonal Balance**: Overuse of oxytocin agonists could potentially interfere with the body's own natural **oxytocin regulation**. The body's feedback loop may adapt to the external supply, causing dysregulation of normal oxytocin production.

- **Impact on Social and Emotional Processing**: Chronic use of oxytocin agonists could alter natural emotional responses, particularly in areas such as **trust**, **attachment**, and **empathy**. While these are typically beneficial traits, there are concerns that **artificially-enhanced** oxytocin levels may skew or interfere with genuine emotional processing.

- **Immunosuppression**: Some animal studies suggest that chronic administration of oxytocin may lead to a suppression of certain immune responses, making the body more vulnerable to infections or immune-related conditions.

- **Increased Risk of Cardiovascular Problems**: There is a small risk of elevated blood pressure and **tachycardia** with prolonged use of oxytocin agonists, particularly in individuals with pre-existing heart conditions.

Long–Term Risk Mitigation:

To mitigate long-term risks, it is essential for healthcare providers to use **oxytocin agonists** judiciously, limiting their duration and frequency of use. A **gradual tapering** of dosages and frequent monitoring of **blood pressure**, **heart rate**, and **kidney function** are crucial for ensuring the safety of patients undergoing long-term oxytocin therapy. Additionally, **psychological support** should be considered to ensure that the benefits of oxytocin agonist treatment do not mask underlying psychological issues, leading to long-term dependency.

3. Contraindications for Use of Oxytocin Agonists

Oxytocin agonists are contraindicated in certain **clinical scenarios**, particularly in patients who have specific pre-existing conditions or risk factors. Before administering oxytocin or its derivatives, healthcare providers should carefully assess the **patient's medical history** to ensure its safety.

Key Contraindications:

- **Severe Hypertension**: Oxytocin can raise blood pressure, making its use risky in patients with **pre-existing hypertension** or **cardiovascular disease**. In such cases, the use of oxytocin should be avoided or carefully monitored.

- **Previous Uterine Surgery**: Women with a history of **cesarean sections** or **uterine rupture** are at an increased risk of **uterine rupture** during labor induction with oxytocin, particularly at higher doses.

- **Severe Pre-eclampsia or Eclampsia**: These conditions, characterized by high blood pressure and organ damage, may be aggravated by the vasoconstrictor effects of oxytocin.

- **Multiple Pregnancy**: In cases of **twin pregnancies** or higher-order multiples, the risks of overstimulation of the uterus with oxytocin may lead to premature labor or fetal distress.

- **Placenta Previa or Abnormal Placental Attachments**: In women with these conditions, oxytocin administration can increase the risk of **placental abruption** or hemorrhage.

In cases where oxytocin is contraindicated, healthcare providers may turn to **alternative methods** for managing labor, anxiety, or other conditions that typically benefit from oxytocin therapy.

4. Handling Overdose and Adverse Reactions

While oxytocin is generally safe in therapeutic doses, overdose or misuse can lead to serious complications. Understanding how to manage an overdose of oxytocin agonists is essential for any clinician using these drugs.

Symptoms of Oxytocin Overdose:

- **Prolonged labor** or **excessive uterine contractions** that lead to **fetal distress** or **uterine rupture**.
- **Severe hypotension** or **shock**, particularly if the oxytocin affects the cardiovascular system excessively.
- **Severe water retention**, leading to **hyponatremia**, confusion, seizures, and **respiratory distress**.

Immediate Management of Overdose:

- **Discontinuation of the drug**: The first step in managing an overdose is to immediately stop the infusion or administration of oxytocin.

- **Symptom management**: For **water intoxication**, intravenous administration of **hypertonic saline** may be required to correct electrolyte imbalances.

- **Monitoring and support**: Continuous monitoring of **vital signs**, **urine output**, and **electrolyte levels** is necessary until the symptoms resolve. If uterine overstimulation occurs, medications such as **terbutaline** may be administered to counteract excessive contractions.

- **Emergency care**: In severe cases of overdose, **intensive care** may be required to address shock, seizures, or other life-threatening complications.

5. Conclusion

Oxytocin agonists are an incredibly promising therapeutic tool, offering vast benefits in the treatment of a wide range of medical conditions. However, as with any potent compound, it is essential to understand the **risks** associated with their use. By being aware of potential **side effects**, **contraindications**, and proper **overdose management**, healthcare providers can ensure the safe and effective use of oxytocin agonists in their patients.

Proper monitoring, appropriate dosage adjustments, and careful patient selection are critical to minimizing risks and maximizing the therapeutic benefits of oxytocin agonists. By doing so, we can harness the full potential of these powerful hormones, while keeping patient safety as the highest priority.

Chapter 22: Oxytocin Agonists and Public Health Policy

As interest in the therapeutic applications of oxytocin agonists grows, so too does the importance of understanding the role these substances play in **public health policy**. Governments, regulatory bodies, and healthcare systems must navigate the complexities of ensuring the safe, equitable, and ethical distribution and use of oxytocin-related drugs. From **labor induction** to **mental health treatments**, oxytocin agonists offer profound benefits, but they also raise significant questions about their regulation, accessibility, and societal impact.

This chapter will explore the landscape of **public health policy** as it pertains to oxytocin agonists, considering issues such as **drug regulation**, **accessibility**, **equity in healthcare**, and the role of **public health initiatives** in advancing research and application.

1. Regulation of Oxytocin Agonists

The regulation of oxytocin agonists is a critical aspect of ensuring their safe and appropriate use in both clinical and non-clinical settings. Regulatory agencies such as the **Food and Drug Administration (FDA)**, the **European Medicines Agency (EMA)**, and other national bodies have specific guidelines for **drug approval** and **monitoring**. For oxytocin agonists, these guidelines must balance the potential therapeutic benefits with safety concerns, particularly when the drugs are used outside traditional reproductive contexts.

Regulatory Challenges:

- **New Applications**: While oxytocin is already well-regulated for use in **childbirth** and **postpartum care**, its expanding role in areas such as **mental health** and **addiction treatment** presents new regulatory challenges. These include determining appropriate dosages, assessing the long-term safety of non-reproductive uses, and ensuring that oxytocin agonists are used ethically in therapeutic settings.

- **Off-Label Use**: The potential for **off-label use** of oxytocin agonists, particularly for emotional or psychological conditions, raises concerns about the **unmonitored distribution** of these substances. Regulating off-label prescriptions becomes complex, especially when the evidence for efficacy in new indications is still evolving.

- **Safety and Monitoring**: Continuous monitoring of side effects, long-term impacts, and drug interactions is necessary. Regulatory bodies may need to implement new systems for tracking the effects of these drugs, including **post-market surveillance**.

Policies for Safe Use:

- **Clear Guidelines for Healthcare Providers**: Regulatory agencies must develop clear guidelines for healthcare providers regarding **dosage**, **administration**, and **patient selection** for oxytocin agonist treatments. Regular education and updates for practitioners can reduce the risk of misuse or adverse reactions.

- **Adverse Event Reporting Systems**: Systems to report adverse events or side effects of oxytocin agonist therapies should be in place to inform ongoing regulatory decisions and updates to safety guidelines.

2. Access to Oxytocin Agonists

Ensuring **equitable access** to oxytocin agonists, particularly for vulnerable populations, is a central issue in public health policy. Access to these medications is influenced by factors such as **cost**, **availability**, **insurance coverage**, and **geographic location**.

Global Disparities:

- **Developed vs. Developing Countries**: In high-income countries, oxytocin agonists are typically available for approved uses like **labor induction** and **postpartum hemorrhage control**. However, in low- and middle-income countries, **access to quality medications** may be limited. The **cost of treatment** and the availability of **trained healthcare providers** can restrict access, leading to disparities in maternal and child health outcomes.
- **Non-Reproductive Uses**: The **off-label use** of oxytocin agonists for mental health disorders, pain management, and other therapeutic applications adds a layer of complexity. These uses may not be covered by insurance in some regions, further restricting access for individuals who could benefit from these treatments.

Public Health Solutions:

- **Subsidized Access in Low-Income Countries**: Governments and international organizations can work together to ensure that **essential medications** like oxytocin agonists are made available in underserved regions. This might involve subsidized pricing, bulk purchasing agreements, or international cooperation.

- **Insurance Coverage Expansion**: In high-income countries, advocating for **expanded insurance coverage** for non-reproductive uses of oxytocin agonists can help increase access to treatment for individuals with conditions such as **anxiety**, **autism spectrum disorder**, or **addiction**.

- **Global Research Initiatives**: Public health agencies could sponsor or support international research efforts to better understand the broader therapeutic potential of oxytocin agonists, and how to optimize their use across diverse populations.

3. Ethical Implications of Oxytocin Agonists

As oxytocin agonists move beyond traditional reproductive uses, ethical considerations will continue to evolve. The manipulation of oxytocin levels in the body to influence **emotions, social behavior**, or **attachment** raises significant ethical questions about the **boundaries of consent, autonomy**, and the potential for **misuse**.

Ethical Concerns:

- **Manipulation of Emotional States**: The ability to influence emotional bonding and attachment with oxytocin agonists could lead to ethical dilemmas, particularly in non-medical contexts (e.g., manipulating social behavior, romantic relationships, or group dynamics). Policies must ensure that the use of oxytocin agonists for these purposes is **voluntary**, **consensual**, and **ethically sound**.

- **Risk of Overuse**: The long-term use of oxytocin agonists for non-reproductive purposes, such as **anxiety management** or **behavioral therapy**, raises concerns about **dependency** or the **pathologizing** of normal emotional responses. There may be an ethical obligation to balance pharmacological treatments with **psychosocial support** to avoid over-reliance on pharmaceutical solutions.

Ethical Frameworks:

- **Informed Consent**: Healthcare providers should ensure that patients receiving oxytocin agonists for non-reproductive purposes are fully informed about the potential risks and benefits, as well as the **long-term implications** of treatment.

- **Regulating Marketing and Access**: There should be strict oversight of the marketing and distribution of oxytocin agonists to prevent **misleading advertising** or **exploitation** of vulnerable populations for commercial gain. Ethical marketing should be based on **evidence-based benefits** and **realistic outcomes**.

- **Universal Ethics Committees**: Establishing independent **ethics review boards** that specifically address the emerging use of oxytocin agonists in **mental health**, **social behavior**, and **psychotherapy** could guide responsible policy and practice.

4. Public Health Initiatives and Research

As the understanding of oxytocin agonists expands, public health agencies must take a proactive role in **funding** and **supporting research** into their broader applications. These agencies can also spearhead **public health initiatives** that raise awareness of the potential benefits and risks associated with oxytocin agonists.

Research Funding:

- **Investing in Innovative Research**: Governments can allocate funding to **multidisciplinary research** exploring the potential of oxytocin agonists in **neurology, psychiatry, pain management**, and **aging**. Increased research could lead to better understanding, optimized dosing regimens, and new therapeutic areas.

- **Collaboration with Universities and Institutes**: Collaborations with academic institutions and pharmaceutical companies can accelerate research into the **long-term impacts** and **synergistic effects** of oxytocin agonists when used alongside other medications.

Public Awareness Campaigns:

- **Educating Healthcare Professionals and the Public**: Public health campaigns should focus on educating healthcare professionals about the benefits and risks of oxytocin agonists, especially in **off-label applications**. This ensures that healthcare providers are equipped to offer informed recommendations to patients.

- **Ethical Public Discussions**: As new uses for oxytocin agonists emerge, it is vital for policymakers to engage in **public discussions** about the ethical, social, and medical implications. Transparent, open conversations can help ensure that public health policies are aligned with the values of society.

5. Conclusion

As oxytocin agonists move into the spotlight of modern medicine, it is essential that **public health policy** evolves to meet the challenges and opportunities they present. From **regulating their use** to ensuring **equitable access**, and from navigating **ethical considerations** to fostering **innovative research**, public health agencies and policymakers must work together to guide the responsible integration of oxytocin agonists into global healthcare systems. By doing so, they can ensure that these powerful substances are used safely, ethically, and effectively, for the betterment of public health worldwide.

Chapter 23: The Intersection of Oxytocin and Other Therapeutic Agents

The therapeutic potential of **oxytocin agonists** is vast, extending far beyond their traditional applications in reproductive medicine. As research progresses, there is growing interest in combining oxytocin with other **neuroactive agents** to amplify its effects, address complex conditions, and optimize patient outcomes. This chapter explores the emerging field of **combination therapies**, where oxytocin is used alongside other drugs to harness synergistic effects in both psychiatric and physical health.

1. Understanding Combination Therapies

Combination therapy refers to the simultaneous use of two or more drugs to achieve a therapeutic effect that is greater than the effect of each drug used alone. This approach has been widely used in the treatment of chronic diseases like **HIV, cancer**, and **hypertension**, but its application in **neuropsychiatric** and **behavioral disorders** is relatively new. When it comes to **oxytocin agonists**, combining them with other **neuroactive** or **endocrine-modulating drugs** holds great promise for improving the **efficacy** and **safety** of treatment protocols.

The **interaction** between oxytocin and other drugs can occur through several mechanisms, including:

- **Synergistic effects**: When the combined effects of two drugs are greater than the sum of their individual effects.
- **Additive effects**: When the combined effects of two drugs are simply the sum of their individual effects, but still result in enhanced therapeutic outcomes.
- **Antagonistic effects**: Where one drug counteracts or diminishes the effect of the other. This can be beneficial in some contexts, such as managing side effects or modulating the potency of oxytocin agonists.

2. Combining Oxytocin Agonists with Psychiatric Medications

Oxytocin's profound impact on **social cognition**, **emotional regulation**, and **interpersonal bonding** has opened up new avenues for combining it with psychiatric medications, particularly in disorders where these functions are disrupted, such as **anxiety**, **depression**, **autism spectrum disorders (ASD)**, and **schizophrenia**.

Oxytocin + Antidepressants

- **Serotonin Reuptake Inhibitors (SSRIs)**: SSRIs, a commonly prescribed class of antidepressants, work by increasing the availability of **serotonin** in the brain, improving mood and emotional regulation. There is evidence to suggest that **oxytocin** might enhance the effects of SSRIs, particularly in terms of improving **empathy**, **social connection**, and **stress resilience**. Research shows that oxytocin administration can help mitigate the emotional blunting often seen in patients taking SSRIs, creating a more holistic approach to treating **major depressive disorder**.

- **Tricyclic Antidepressants (TCAs)**: TCAs also influence neurotransmitters like **norepinephrine** and **serotonin**, and studies suggest that oxytocin could work in synergy with these agents to enhance mood regulation and **reduce social withdrawal**—a key symptom of depression.

Oxytocin + Antipsychotics

- **Atypical Antipsychotics**: Medications such as **clozapine** and **risperidone** are commonly used to manage symptoms of **schizophrenia** and other psychotic disorders. Oxytocin agonists have been shown to potentially enhance **cognitive function, social interaction**, and **emotional regulation** in patients with psychotic disorders. By combining oxytocin with antipsychotic drugs, clinicians might be able to improve outcomes, particularly in managing social deficits and enhancing interpersonal functioning.

- **Cognitive Behavioral Therapy (CBT)**: Combining oxytocin with psychotherapy such as CBT could potentially strengthen therapeutic engagement, improving both the emotional and cognitive aspects of therapy for individuals with **schizophrenia** or **bipolar disorder**.

3. Oxytocin and Pain Management

Oxytocin's well-documented **analgesic properties** make it a valuable adjunct to pain management, particularly in the context of **chronic pain, post-surgical recovery**, and **neuropathic pain**. The combination of oxytocin agonists with traditional pain relievers like **opioids, non-steroidal anti-inflammatory drugs (NSAIDs)**, or **anticonvulsants** could open up new pathways for more effective pain control.

Oxytocin + Opioids

While opioids have been the standard for managing severe pain, they come with a high risk of **dependency** and **tolerance**. Combining oxytocin with opioids may allow for lower doses of opioids, potentially reducing the risk of addiction and overdose. Some research indicates that oxytocin may help to counteract some of the **negative emotional effects** of opioid use, including **stress** and **anxiety**, creating a more balanced pain management regimen.

Oxytocin + NSAIDs

NSAIDs, such as **ibuprofen** or **aspirin**, work by inhibiting the production of **prostaglandins**, which are chemicals involved in the inflammatory process. Oxytocin's ability to modulate **inflammation** and reduce **stress-induced pain** could complement the effects of NSAIDs, providing more comprehensive pain relief in conditions like **osteoarthritis** or **muscle injuries**.

4. Oxytocin and Hormonal Therapies

Oxytocin interacts with several other hormones, and combining it with agents that influence **hormonal balance** could have significant therapeutic benefits. For instance, pairing oxytocin with **hormonal contraceptives** or **hormone replacement therapy (HRT)** could enhance **mood regulation** and **social bonding** in individuals undergoing hormonal treatments.

Oxytocin + Estrogen

Estrogen has long been associated with **emotional regulation** and **bonding behavior**. Some studies suggest that **estrogen** may amplify the **effects of oxytocin** on the brain, potentially improving **mood** and **social cognition**. This combination could be particularly useful in treating conditions like **postpartum depression, anxiety**, or **menopausal symptoms**.

Oxytocin + Cortisol Modulators

Cortisol, the body's primary **stress hormone**, often interacts with oxytocin to influence social and emotional behavior. Combining oxytocin agonists with **cortisol-lowering agents** like **DHEA** or **adaptogens** could lead to more effective treatments for stress-related disorders such as **chronic fatigue syndrome**, PTSD, or **generalized anxiety disorder**.

5. The Role of Oxytocin in Multimodal Treatment Approaches

While combination therapies typically involve the use of **pharmaceutical drugs**, oxytocin's effects on **emotion**, **social bonding**, and **behavior** suggest that it may be particularly effective when integrated into **multimodal treatment plans**. For example, combining oxytocin agonists with **psychotherapeutic approaches** (e.g., **cognitive-behavioral therapy (CBT)**, **dialectical behavior therapy (DBT)**), **mindfulness practices**, and **physical therapy** could address a wide range of psychiatric and physical conditions in a holistic manner.

Case Studies:

- **Autism Spectrum Disorder (ASD)**: Oxytocin has been investigated as part of a multimodal treatment for ASD. Combining oxytocin with **behavioral therapy** or **occupational therapy** has shown promise in improving **social interaction**, **empathy**, and **communication skills** in children with ASD.
- **Trauma Recovery**: In individuals recovering from **trauma** or **PTSD**, combining oxytocin with **somatic therapies** (e.g., **EMDR (Eye Movement Desensitization and Reprocessing)** or **sensorimotor psychotherapy**) could facilitate emotional processing and recovery while improving the **trust** and **attachment** aspects of therapy.

6. Challenges in Combining Oxytocin with Other Drugs

While combination therapies offer great promise, there are several challenges that need to be addressed:

- **Drug Interactions**: The potential for adverse drug interactions between oxytocin and other therapeutic agents must be thoroughly studied. For example, combining oxytocin with certain **antidepressants** or **antipsychotics** could have unpredictable effects on mood or **cognitive function**.

- **Optimal Dosage**: Determining the right dosage of oxytocin when used in combination with other drugs is critical. Too much oxytocin could lead to **overstimulation** of receptors, while too little may not have the desired effects. Careful titration and patient monitoring are essential.

- **Long-Term Effects**: While combination therapies might offer immediate benefits, understanding their **long-term effects** is critical for developing sustainable treatment regimens.

7. Conclusion: The Future of Oxytocin Combination Therapies

The future of **oxytocin agonists** in combination therapies holds tremendous potential for enhancing treatment outcomes across a range of conditions. As research continues, we can expect more **personalized** and **multidisciplinary approaches** to emerge, integrating oxytocin with other neuroactive agents, hormones, and therapeutic practices. By understanding the complex interactions between oxytocin and other drugs, healthcare providers can optimize treatment plans, improving both the **efficacy** and **safety** of interventions, and ultimately helping patients achieve better long-term outcomes in both mental and physical health.

Chapter 25: Conclusion and Looking Forward

As we draw this comprehensive guide on **oxytocin agonists** to a close, it's essential to reflect on the remarkable journey we've taken through the **science**, **applications**, and **therapeutic potential** of this fascinating hormone. Oxytocin, often called the "love hormone," has proven itself to be much more than a molecule tied to childbirth and lactation. From its pivotal role in **emotional bonding** and **social behavior** to its potential in **mental health** treatments and **pain management**, oxytocin agonists are carving out a path toward a new era in both **medicine** and **human connection**.

Key Insights and Takeaways

1. **Oxytocin's Multifaceted Role in Physiology**: Throughout this book, we've seen the profound influence of oxytocin in **human physiology**—especially in **reproductive health, emotional regulation, bonding,** and **social behavior**. Its wide-reaching effects on the **brain, endocrine system,** and **immune function** make it a powerful tool in a range of clinical applications. From **labor induction** and **maternal bonding** to **stress relief** and **anxiety reduction**, oxytocin impacts numerous aspects of health, creating a foundation for its diverse therapeutic uses.

2. **The Potential of Oxytocin Agonists**: Oxytocin agonists—substances that enhance or mimic the action of oxytocin—are at the forefront of many innovative treatments. Whether used alone or in combination with other drugs, these agonists hold promise for enhancing **social connections, emotional regulation, mental health treatment,** and **pain management**. As we explored, oxytocin agonists have potential applications in conditions like **autism, anxiety disorders, addiction,** and **chronic pain**, offering a novel approach to previously difficult-to-treat issues.

3. **The Ethical and Safety Considerations**: While oxytocin agonists offer tremendous promise, the ethical questions surrounding their use—especially in terms of emotional manipulation and behavioral modification—must not be overlooked. As with any powerful tool, there are risks of **misuse**, **dependency**, or **unintended consequences**. A careful balance must be struck between the therapeutic potential and the ethical implications of manipulating such a powerful hormone, especially in areas like **relationship counseling**, **parenting**, and **addiction treatment**.

4. **Integration into Holistic and Personalized Care**: Oxytocin agonists are not a standalone solution. Their real potential lies in their integration into **personalized care plans** that consider the unique needs of individuals. Combining oxytocin agonists with therapies like **psychotherapy**, **mindfulness**, and **cognitive behavioral therapy** creates a holistic approach to mental and emotional health. Additionally, the integration of oxytocin in **personalized medicine**—tailored to an individual's genetic makeup and health conditions—holds the promise of optimizing **treatment outcomes** for patients, particularly in areas like **mental health**, **chronic pain**, and **neurodegenerative diseases**.

5. **The Future of Oxytocin Agonists in Medicine**: As research continues, we can expect **exciting advancements** in our understanding of oxytocin and its agonists. From new drug formulations and **delivery methods** to insights into how oxytocin works in the **brain** and **body**, the future is ripe with potential. The expanding field of **combination therapies**—where oxytocin is used alongside other neuroactive agents—may unlock even more powerful treatment protocols for conditions like **depression, schizophrenia, post-traumatic stress disorder (PTSD)**, and **neurodegenerative diseases**.

6. **A Changing Landscape in Public Health**: As more **public health policies** are shaped around the use of oxytocin-related drugs, we will likely see greater **accessibility** and **integration** of these therapies within mainstream healthcare. **Regulatory frameworks** will need to evolve to address the unique challenges posed by oxytocin agonists, particularly concerning **misuse** and **misinterpretation** of their effects in non-clinical settings.

Final Thoughts on the Future of Oxytocin Agonists

The power of oxytocin in shaping **human connection, emotional well-being**, and **physiological health** is just beginning to be fully realized. As science uncovers new insights and the clinical landscape continues to evolve, we are poised on the brink of a transformation in the way we approach **mental health, social behavior**, and **medicine** at large.

Oxytocin agonists could be the cornerstone of a **paradigm shift** in **healthcare**, offering not only a tool for treating physical conditions but also for enhancing **emotional resilience**, **social functioning**, and **psychological well-being**. However, as with any emerging field, the path forward requires ongoing **research**, **ethical consideration**, and **careful application**.

As we look forward, the integration of oxytocin agonists into therapeutic regimens will not only offer **tangible benefits** for individual health but could also contribute to a **more compassionate, connected**, and **emotionally intelligent society. Holistic health, personalized care**, and **innovative drug therapies** are rapidly becoming the foundation of the next era of healthcare, and oxytocin agonists will undoubtedly play a pivotal role in shaping this future.

The Next Era of Healthcare: Oxytocin as a Pivotal Tool

We stand at the threshold of a future where **oxytocin agonists** could become essential components of the **next generation** of healthcare. Their potential is vast, touching nearly every aspect of human experience—from **childbirth** and **parenting** to **mental health** and **aging**. As we continue to unlock the mysteries of oxytocin's complex biological and emotional influence, we must embrace this moment of scientific discovery to create a future where healthcare is **integrated, personalized**, and **holistic**.

The future of **oxytocin agonists** holds boundless promise—not just as a tool for healing the body, but for nurturing the bonds that define the **human experience**.

4o mini

www.ingramcontent.com/pod-product-compliance
Lightning Source LLC
Chambersburg PA
CBHW082106220526
45472CB00009B/2059